CRE公務員綜合招聘考試

公務員招聘

能力傾向

測試精讀王 NOTE

RE各項能力 極速提升

鬆成為薪優糧準公務員

資深大專課程導師編撰

目緊貼CRE形式及深淺

擬試卷及答案詳盡講解

Fong Sir 著

序言

公務員薪高糧準，要成功通過公務員招聘，以學位／專業程度職系而言，最基本的要求就是通過公務員綜合招聘考試（Common Recruitment Examination，簡稱CRE），該測試首先包括三張各為45分鐘的多項選擇題試卷，分別是「中文運用」、「英文運用」、和「能力傾向測試」，其目的是評核考生的中、英語文能力及推理能力。

之後是「基本法測試」試卷，基本法測試同樣是以選擇題形式作答之試卷，全卷合共15題，考生必須於20分鐘內完成。而基本法測試本身並無設定及格分數，滿分則為100分。基本法測試的成績，會對於應徵「學位或專業程度公務員職位」的人士佔其整體表現的一個適當的比重。

然而，一些考生雖有志加入公務員行列，但礙於此一門檻，因而未能加入公務員團隊。

有見及此，本書特為應考公務員綜合招聘試的考生提供試前準備，希望考生能熟習各種題型及答題方法。可是要在45分鐘之內完成全卷對大部分考生而言確有一定的難度。因此，答題的時間分配也是通過該試的關鍵之一。考生宜通過本書的模擬測試，了解自己的強弱所在，從而制訂最適合自己的考試策略。

此外，考生也應明白任何一種能力的培訓，固然不可能一蹴而就，所以宜多加推敲部分附有解說的答案，先從準確入手，再提升答題速度。考生如能善用本書，對於應付公務員綜合招聘考試有很大的幫助。

目錄

題庫練習一

(一)演繹推理

答題指示

請根據以下短文的內容，選出一個或一組推論。請假定短文的內容都是正確的。

例題：

龍門鎮中的每一個人都是亞華的親戚，亞華只有一個兒子，銘希在龍門鎮裡生活，志南是亞華的丈夫。由此可推論：

    A. 銘希是亞華的女兒。

    B. 亞華生活在龍門鎮裡。

    C. 銘希是亞華的親戚。

    D. 志南生活在龍門鎮裡。

答案：C

解析：短文第三句指出銘希在龍門鎮裡生活，而從第一句可知道銘希是亞華的親戚。

練習題：

請根據以下短文的內容，選出一個或一組推論。請假定短文的內容都是正確的。

1. 八位大學學者趙教授、錢教授、孫教授、李教授、王所長、陳博士、周博士、沈局長在爭取一項科研基金。按規定只有一人能獲得該基金。由大學評委投票決定。已知：如果錢教授獲得的票數比周博士多，那麼李教授將獲得該項基金；如果沈局長獲得的票數比孫教授多，或者李教授獲得的票數比王所長多，那麼陳博士將獲得該基金；如果孫教授獲得的票數比沈局長多，同時周博士獲得的票數比錢教授多，那麼趙教授將獲得該項基金。

1. 如果陳博士獲得了該項基金，那麼下面哪個結論一定是正確？

   A. 孫教授獲得的票數比沈局長多

   B. 沈局長獲得的票數比孫教授多

   C. 李教授獲得的票數比王所長多

   D. 錢教授獲得的票數不比周博士多

2. 公文是指國家機關在處理公務的過程中，形成的具有法定效力和規範體式的文字材料。

   根據以上定義，下列不屬於公文的一項是：

   A. 石家莊市人民政府《關於授予滹沱河生態開發整治工程市長特別獎的決定》

PART ONE
題庫練習

PART TWO
模擬試卷

PART THREE
考生急症室

B. 文化部《關於同意邀請新加坡歌手林俊傑到江蘇演出的批復》

C. 北京市政府《關於2007年北京地區普通中等專業學校畢業生畢業派遣工作安排的通知》

D. 東北師範大學教務處《關於2008年度工作情況的報告》

3. 有些墨西哥人不愛吃辣椒。有些愛吃甜食的人不愛吃辣椒。

以下哪項能保證上述推理成立？

A. 所有墨西哥人都不愛吃辣椒

B. 有些墨西哥人愛吃甜食

C. 所有愛吃甜食的人都愛吃辣椒

D. 所有墨西哥人都愛吃甜食

4. 如果A考試及格了，那麼B、C和D肯定也及格了。由此可知：

A. 如果A考試沒及格，那麼B、C和D中至少有一個沒及格

B. 如果B、C和D都及格了，那麼A肯定也及格了

C. 如果D的成績沒有及格，那麼B和C不會都考及格

D. 如果C的成績沒有及格，那麼A和D不會都考及格

5. 如果貫徹絕對公平，那麼必然導致按勞分配；若按勞分配，將出現貧富不均；只有貧富均等，才能貫徹絕對公平。所以：

A. 必須實行按勞分配

B. 必須實行按需分配

C. 必須貫徹絕對公平

D. 不能貫徹絕對公平

6. 「皮膚中膠原蛋白的含量決定皮膚是否光滑細膩，決定人的皮膚是否年輕。相同年齡的男性和女性皮膚中含有相同量的膠原蛋白，而且女性更善於保養，並能從日常保養中提高皮膚膠原蛋白含量，盡管如此，女性卻比男性更容易衰老。」以下選項能解釋上述矛盾的是：

A. 男性皮膚內膠原蛋白是網狀結構，而女性是絲狀結構的

B. 女性維持光滑的皮膚，年輕美貌的容顏需大量膠原蛋白

C. 只有蹄筋類食物富含膠原蛋白，但很難被人體消化吸收

D. 男性膠原蛋白幾乎不消耗，女性代謝需耗大量膠原蛋白

PART ONE
題庫練習

PART TWO
模擬試卷

PART THREE
考生急症室

7. 甲、乙、丙、丁四位同學在一起議論本班參加某項活動情況：

甲說：我班所有同學都參加了。

乙說：如果小明沒參加，那麼小華也沒參加。

丙說：小華參加了。

丁說：我班所有同學都沒有參加。

已知四人中只有一人說的不正確，由此可見：

A. 甲說的不正確，小明沒參加

B. 乙說的不正確，小明參加了

C. 丙說的不正確，小明沒參加

D. 丁說的不正確，小明參加了

8. 以下關於A電腦故障的陳述中，只有一個是真的。這一真的判斷是：

A. 顯示卡壞了

B. 主板壞了，那麼記憶體也一定出現了故障

C. 主板或顯示卡壞了

D. 主板壞了

9. 近來東南亞局勢不穩定，是由於禽流感的影響。但假如沒有經濟形勢的動蕩不安，則禽流感對局勢的影響將不明顯。因此，為了阻止局勢不穩定必須穩定經濟形勢。

以下哪項如果為真，則最能削弱上述結論？

A. 經濟形勢的動蕩不安與禽流感沒有關係

B. 在禽流感之前，東南亞局勢已經不穩定

C. 東南亞沒有嘗試預防禽流感

D. 在經濟形勢動蕩之前，類似禽流感之類的疫情已多次造成東南亞的不穩定

10. 關於一個班的英語六級通過情況有如下陳述：

（1）班長通過了

（2）該班所有人都通過了

（3）有些人通過了

（4）有些人沒有通過

經過詳細調查，發現上述斷定只有兩個是正確的。可見：

A. 該班有人通過了，但也有人沒有通過

B. 班長通過了

C. 所有人都通過了

D. 所有人都沒有通過

PART ONE
題庫練習

PART TWO
模擬試卷

PART THREE
考生急症室

11. 下面是甲、乙、丙三位面試老師關於錄取研究生的意見：

甲：如果不錄取方先生，那麼不錄取王先生

乙：如果不錄取王先生，那麼錄取方先生

丙：如果錄取方先生，那麼不錄取王先生

應該選擇何種錄取方案，使甲、乙、丙三位面試老師的要求同時得到滿足。

A. 只錄取王先生

B. 只錄取方先生

C. 王先生、方先生都錄取

D. 王先生、方先生都不錄取

12. 通過實際的調查發現，有的被告是無辜的，甚至是被誣陷的；並非所有的嫌疑人都是罪犯。

如果上述調查的情況是真實的，那麼以下哪項一定為真？

A. 有很多嫌疑人是被誣陷的

B. 所有的嫌疑人都不是罪犯

C. 有的嫌疑人不是罪犯

D. 大多數嫌疑人都是罪犯

13. 某大學哲學系邏輯學專業共有30名本科生，男女各一半，其中20人喜歡公理集合論，25人喜歡模型論。那麼以下哪項是不可能的，除了：

    A. 10個男生喜歡公理集合論而不喜歡模型論

    B. 10個喜歡模型論的男生不喜歡公理集合淪

    C. 15個喜歡模型論的女生不喜歡公理集合論

    D. 15個喜歡公理集合論的男生只有5個喜歡模型論。

14. 某辦公室共有3人：主任1人，副主任1人，辦事員1人。

    （1）主任懂日語

    （2）有人不懂日語

    （3）有人懂日語

    在上述三個判斷中只有一個是真的，由此看見：

    A. 副主任懂日語

    B. 副主任不懂日語

    C. 主任懂日語

    D. 主任不懂日語但辦事員懂日語

15. 不確定性避免是指在任何一個社會中，人們對於不確定的、含糊的、前途未卜的情景，都會感到面對的是一種威脅，從

PART ONE
題庫練習

PART TWO
模擬試卷

PART THREE
考生急症室

而總是試圖加以防止。

根據上述定義，下列不屬於不確定性避免的是：

A. 學校號召研究生出過深造，王同學覺得自己生活比較安定，不想出去，就說：「我的英語水平不行，還是把機會讓給其他同學吧！」

B. 隨著金融風暴的到來，股市漲跌起伏不定，趙先生將自己手上的股票迅速拋掉。

C. 畢業答辯日期臨近，李先生對於答辯老師可能提出的問題心中沒底，非常緊張。

D. 張先生在去外資公司工作還是自己創業之間反覆思考，覺得後者風險太大，於是選擇進入外資公司工作。

16. 文化反哺是指在急速的文化變遷時代所發生的年長一代向年輕一代進行廣泛的文化吸收的過程。

根據上訴定義，下列屬於文化反哺的是：

A. 七十年代出生的人並不是狂熱的追星族，但很多人都非常喜歡八十年代出生的歌星創造的一些歌曲。

B. 張老師說：「現在教學都是電子化，很多時候要做PPT，要上網查資料，遇到不會的時候，兒子就是我的老師。」

C. 孫經理說，他們公司年輕人居多，他嘗試著用跟兒子交流的方式與年輕人溝通，果然效果很好。

D. 劉主任接受了下屬們的建議，制定了新的工作制度，提高了員工的工作積極性。

17. 社會促進是指個人從事某項活動時，他人在場促進其活動完成，提高其活動效率的現象，也稱社會助長。

根據上述定義，下列不屬於社會促進的是：

A. 學生在進行數學計算時，教師的旁觀可使學生充分開動腦筋。

B. 王先生扶起跌倒的李老伯，他覺得自己做了一件有意義的事。

C. 員工在進行生產時，廠長站在車間裡，員工效率就會大大提升。

D. 足球運動員走進球場時，看到比賽場座無虛席就會鬥志昂揚。

18. 庭院經濟是指農戶充分利用家庭庭落的空間、周圍非承包的空坪隙地和各種資源，從事高度集約化商品生產的一種經營形式，主要有種植業、養殖業、加工業。

根據上述定義，下列屬於庭院經濟的是：

A. 在陽台上種植花草美化居住環境

B. 在承包的池塘養殖鯉魚

C. 在自家後院加蓋瓦房建立手工藝編織廠

D. 在草原上放牧羊群

19. 職業病是指企業、事業單位和個體經濟組織的勞動者在職業活動中，因接觸粉塵、放射性物質和其他有毒、有害物質等因素而引起的疾病。

PART ONE
題庫練習

PART TWO
模擬試卷

PART THREE
考生急症室

根據上述定義，下列屬於職業病的是：

A. 李老師搬進了剛裝修的新辦公室，由於有害氣體嚴重超標，導致她患上了血液病。

B. 刑警王先生在近30年的職業生涯中接觸過各種犯罪分子，在日常生活中，他也經常用職業的眼神打量親朋好友。

C. 趙先生是一家公司的高級管理人員，工作經常加班加點，飲食很不規律，終於被醫院查出患上了胃炎。

D. 張先生在一家皮鞋廠工作，因長期接觸含苯物質而患上了再障性貧血。

20. 股東代表訴訟是指當公司的合法權益受到不法侵害而公司卻息於起訴時，公司的股東即以自己的名義起訴，所獲賠償歸於公司的一種訴訟制度。

根據上訴定義，以下情形可以提起股東代表訴訟的是：

A. 甲公司連續5年盈利，卻不向股東分配利潤。

B. 乙公司由於經營不善，連年虧損，公司股東要求查詢會計賬簿，被公司拒絕。

C. 丙公司經理私自以公司資產為他人提供擔保，公司董事會就此提出訴訟。

D. 某公司公司拖欠丁公司大筆貸款已近兩年，而丁公司董事會和經理未採取任何措施索要。

21. 水平一體化物流是指同一行業的多個企業，通過共同利用物流渠道，獲得規模經濟效益、提高物流效率。水平一體化物流須具備物流需求和物流供應的信息平台，要有大量企業參與並存在較多的商品量。

根據上述定義，下列選項屬於水平一體化物流的是：

A. 某家具廠要求原材料供應商和產品銷售商使用同一家物流公司

B. 某市屠宰企業將豬、牛肉混放入冷藏車並送往一百多家零售店

C. 電子城百家商店簽訂協議，指定其中一家承擔電子城送貨業務

D. 某百貨公司設立物流中心，統籌安排全公司各類貨物發送工作

22. 想像是指在原有經驗的基礎上創造新形像的思維活動。按照想像是否受意志控制，可分為隨意想像和不隨意想像。不隨意想像的特點是把各種印象和信息離奇、突然、有時是無意義地組合在一起。隨意想像是把各種印象和信息自覺控制、有目的、經過意志的努力呈現出需要的場景。

根據上述定義，下列選項屬於隨意想像的是：

A. 李先生昨天晚上睡覺時，夢到了兒時一起嬉戲的伙伴

B. 張先生接到錄取通知，想到自己實現了目標，很興奮

PART ONE
題庫練習
PART TWO
模擬試卷
PART THREE
考生急症室

C. 王先生面對設計圖，憧憬著公司新大樓竣工後的樣子

D. 陳先生的父親看著照片回憶起當年上山下鄉對的場景

23. 虛擬人力資源管理是指以合作關係為基礎，充分利用現代信息網絡技術，幫助企業獲取、發展和籌劃智力和勞力資本的一種人力資源管理辦法，它可以滿足企業管理虛擬化發展的要求，將大量的人力資源管理活動外部化或由員工實現自主管理，企業從而可以將主要精力放在核心人力資源管理萬面，提高人力資源管理效率。

根據上述定義，下列選項不屬於虛擬人力資源管理的是：

A. 某企業通過獵頭公司物色到一位產品研發專家

B. 某集團將其全部廣告交由同一家廣告公司設計

C. 李先生和同事都被公司派到某職業學校培訓半年

D. 某公司委托勞動派遣公司負責分發員工的薪酬

24. 網絡暴力是行為主體的網絡行為對當事人造成實質性傷害的網絡失範現象。

根據上述定義，下列選項屬於網絡暴力的是：

A. 某非法傳銷團伙借助網絡發展下線

B. 某足球俱樂部球迷建立網站，要求領隊被辭退

C. 某藝人喜歡在網上爆料，提高自己網絡的訪問量

D. 某選秀選手隱私在網上被曝光後，宣布放棄晉級機會

25. 積極錯覺是指當自我由於消極的信息而使自尊心面臨威脅時，用理想化的自我、不現實的樂觀或誇大的可控性感知等作為緩衝器，來保護自己的自尊。根據上述定義，下列選項屬於積極錯覺的是：

A. 某男生追求女生失敗、認為對方是因為害羞不敢答應

B. 很多人覺得考試時間很緊張，可林先生還覺得時間足夠

C. 某人擔心經理對自己印象差，每天提早到辦公室上班

D. 某男研究生屢次失敗，但其堅信天道酬勤，始終會成功

26. 甲、乙二人均為木材廠的工人。某日為某房主搬運木材，休息時，甲說不知這木材是否能燃燒，乙說我去試試，說完乙便用打火機去點，結果引燃了旁邊的油桶，將房主的房子燒毀。乙的行為屬於：

A. 疏忽大意的過失

B. 間接故意

C. 意外事件

D. 過於自信的過失

PART ONE
題庫練習

PART TWO
模擬試卷

PART THREE
考生急症室

27. 在打獵季節，在行人路上行走時被汽車撞傷的人數是樹林中的打獵事故中受傷人數的2倍。因此，在打獵季節，人們在樹林裡比在行人路上行走安全。

要評價上面的論證，最重要的是要知道：

A. 打獵季節在樹林中行走的人數

B. 在打獵季節，馬路上的行人和樹林中人數的比例

C. 打獵季節在人行道行走時被撞傷的人數和總人數的比例

D. 在打獵季節，汽車司機和打獵人都能小心點，那麼受傷的人數將會有所下降

28. 單位犯罪：指公司、企業、事業單位、機關、團體，為本單位謀取利益，經單位決策機構或負責人員決定而以單位名義實施的危害社會、依法應受刑罰處罰的行為。

以下屬於單位犯罪的是：

A. 某牛奶生產企業負責人決定將過期牛奶摻進鮮牛奶中包裝出售，獲利五十多萬元，後發生食物中毒事件被人揭發出來

B. 某企業財務偽造公司印章簽章，欠下巨額債務之後準備逃逸，最後鋃鐺入獄

C. 某公司負責人李先生以公司名義，將當年巨額年終利潤以紅利方式分發給下屬

D. 某公司業務員陳先生冒充公司總經理名義實施詐騙，並在一個月內將十多萬元贓款揮霍一光

29. 根據對2020年劍牛出版社申報的引進國外圖書的選題分析，總計引進選題6737種，其中自然科學類佔13.65%，經濟類佔14.36%。

從這段文字可以推出：

A. 那些雖有學術價值但會有市場風險的國外圖書，往往受到出版商的冷遇

B. 對外國先進科技書引進的比例明顯偏低，人文社會科學引進數量不僅偏大，而且偏重某一類

C. 外國文學古典名著，因為無需買版權，市場一直有穩定的讀者群，因而被大量重複出版

D. 自然科學類圖書的引進，翻譯質量要求比較高、周期比較長，致使翻譯出版的比例比較低

30. 據世界衛生組織估計，目前全球患抑鬱症的人多達1.2億，幾乎每4人中便有1人會在一生中某個階段出現精神或行為問題。到2050年，抑鬱症將位居全球疾病排行榜第二位，僅次於心臟病。非洲目前約有2,600萬人患有不同程度的抑鬱症，不過，與抑鬱症的高發病率形成鮮明對比的是，90%的抑鬱症患者因沒有意識到自己可能患有抑鬱症而未能及時就醫。

根據這段文字，可以知道：

A. 全球患抑鬱症的人多達總人口的1/4

B. 目前心臟病居全球疾病排行榜第一位

PART ONE
題庫練習

PART TWO
模擬試卷

PART THREE
考生急症室

C. 不能及時就醫是造成抑鬱症發病率高的原因

D. 非洲抑鬱症患者中及時就醫者不超過300萬

31. 美國著名的經濟學家阿瑟‧奧肯發現了周期波動中經濟增長率和失業率之間的經驗關係，即當實際GDP增長相對於潛在GDP增長（美國一般將之定義為3%）下降2%時，失業率上升大約1%；這條經驗法則以其發現者為名，稱為「奧肯定律」。潛在GDP這個概念是奧肯首先提出的，它是指在保持價格相對穩定情況下，一個國家經濟所產生的最大產值。潛在GDP也稱充分就業GDP。

下列和上述的分析方法相同的是：

A. 青蛙的雙腿對水面的蹬力很大，水給青蛙的反作用力也很大，人們從中受到啟示，模仿青蛙發明了蛙泳的游泳方式

B. 齒輪的轉速和齒數成反比，即齒輪的齒數越多，齒輪轉速越慢

C. 蝙蝠在夜間能夠快速飛行，可以躲避障礙物，這是因為它的耳朵能聽到超聲波遇到障礙物時產生的回聲。如果把蝙蝠的耳朵堵上，它就會撞到障礙物；如果把蝙蝠的眼睛蒙上，它卻不會撞到障礙物

D. 天上有了掃帚雲，不出三天大雨淋

32. 陳小姐、李小姐、楊小姐、朱小姐和吳小姐是同一家公司的同事，因工作的需要，她們不能同時出席公司舉辦的新產品發布會。她們的出席情況是：

（1）只有陳小姐出席；李小姐、楊小姐和朱小姐才都出席

（2）陳小姐不能出席

（3）如果李小姐不出席，楊小姐也不出席

（4）如果楊小姐不出席，吳小姐也不出席

（5）已經決定吳小姐出席發布會

根據上述情況，可以推出：

A. 李小姐出席發布會，楊小姐和朱小姐不出席發布會

B. 朱小姐出席發布會，李小姐和楊小姐不出席發布會

C. 楊小姐和朱小姐出席發布會，李小姐不出席發布會

D. 李小姐和楊小姐出席發布會，朱小姐不出席發布會

PART ONE
題庫練習

PART TWO
模擬試卷

PART THREE
考生急症室

33. 甲、乙、丙和丁四人的國籍分別是：英國、俄國、法國、日本。乙比甲高，丙最矮；英國人比俄國人高，法國人最高；日本人比丁高。

   這四個人的國籍是：

   A. 甲是英國人，乙是法國人，丙是俄國人，丁是日本人

   B. 甲是法國人，乙是日本人，丙是俄國人，丁是英國人

   C. 甲是俄國人，乙是法國人，丙是英國人，丁是俄國人

   D. 甲是日本人，乙是法國人，丙是俄國人，丁是英國人

34. 公務員權利是指法律對公務員在履行職責、執行國家公務的過程中可以做出一定行為、要求他人作出某種行為或者抑制一定行為的許可和保障。

   根據以上定義，下列不屬於公務員權利的是：

   A. 公務員全心全意為人民服務

   B. 公務員獲得工資報酬，享受福利、保險待遇

   C. 公務員獲得履行職責應當具有的工作條件

   D. 公務員對機關工作和領導人員提出的批評和建議

答題及解析：

1. D

解析：根據「如果錢教授獲得的票數比周博士多，那麼李教授將獲得該項基金」，而事實為陳博士獲得了該項基金，因只有一個人能獲該項基金，所以李教授未獲得該項基金，根據充分條件假設命題的推理規則，「否定後件則否定前件」，可得錢教授獲得的票數不比周博士多。

2. D

解析：公文的發布主體是國家機關。D項，東北師範大學教務處不屬於國家機關，《關於2008年度工作情況的報告》也不具有法定效力。

3. D

解析：要保證上述推理成立，則先假設以下四項為真，看能否推出題幹的推理。

A項中，所有墨西哥人都不愛吃辣椒與愛吃甜食的人之間並無直接關係，排除；

B項中，有的墨西哥人→愛吃甜食，與有的愛吃甜食的→不愛吃辣椒，不能形成有效遞推關係，所以不能推出有的墨西哥人→不愛吃辣椒，排除；

C項中，所有愛吃甜食的人都愛吃辣椒，根據遞推規則，「所有」可以推出「某些」，則可以推出有些愛吃甜食的人也喜歡吃辣椒，但不能據此推出有些愛吃甜食的人不愛吃辣椒，排除；

D項中，所有墨西哥人都愛吃甜食，因為有些墨西哥人不愛吃辣椒中的「有些人」包含在所有墨西哥人之中，説明有些甜食的墨西哥人不愛吃辣椒，可以推出。故本題答案為D選項。

4. D

解析：如果C的成績沒有及格，這就否定了充分條件假言命題「如果A考試及格，那麼B，C和D肯定也及格了」。所以可以推出否定的前件，即A考試不及格，所以便可推出A和D不會都及格。

PART ONE
題庫練習

PART TWO
模擬試卷

PART THREE
考生急症室

5. D

解析：相互推出關係如下：（1）貫徹絕對公平→按勞分配；（2）按勞分配→貧富不均；（3）貫徹絕對公平→貧富均等。根據（1）（2）可推出，貫徹絕對公平→貧富不均。這恰與（3）相矛盾。所以不能貫徹絕對公平，選擇D。

6. D

解析：選項A雖然指出男性與女性皮膚內膠原蛋白的結構不同，但並未說明其與「女性比男性更容易衰老」的關係。選項B和C都與題幹矛盾無關，D項則合理解釋了矛盾，故選D。

7. D

解析：甲的話和丁的話是「上反對關係」，上反對關係的兩個命題，要麼一個假，要麼兩個都假，所以不正確的一定在甲和丁之間，因為只有一個是不正確的，這即意味著乙和丙都是正確的。丙（小華參加了）是正確的，這就意味著丁（我班所有同學都沒有參加）是不正確的。

8. B

解析：如果A項「顯示卡壞了」必然推出C項「主板或顯示卡壞了」正確。同理，如果D項「主板壞了」正確必然推出C項「主板或顯示卡壞了」正確。根據題幹，只有一個是真的，所以A項和D項都不能為真，否則就有兩個為真，既然A項、D項不能為真，那麼C項也不能為真，所以只能B項為真，這裡注意B項所說的是一種條件關係，並不說明B項前件「主板壞了」是客觀事實。

9. D

解析：題中結論為經濟形勢不穩是局勢不穩定的重要原因，要想削弱這個結論，可找出造成局勢不穩定的其他原因，即類似禽流感之類的疫情。

10. A

解析：陳述中（2）項如果為真，則（1）項必為真，這與題幹「上述斷定只有兩個是真的」不一致，所以（2）項必為假，又因為（2）項和（4）項為矛盾命題，即「必有一真一假」，（2）項為假，則（4）項必為真。又根據題幹「上述斷定只有兩個是真的」，（2）、（4）一假一真，所以（1）、（3）必有一真一假。顯然，如果（1）真那麼（3）必真，這與命題不符，所以（1）為假，（3）為真。

11. B

解析：運用代入法。A項代入甲得到「錄取方先生」，代入丙得到「不錄取方先生」，顯然矛盾，C項代入丙即可推出矛盾；D項代入乙即可推出矛盾。

12. C

解析：並非所有的嫌疑人都是有罪的，這一命題等值於有的嫌疑人不是有罪的。

13. B

解析：A項不滿足題幹「25人喜歡模型論」，C項不滿足「20人喜歡公理集合論」，D項不滿足「25人喜歡模型論」。

14. B

解析：（2）和（3）「下反對關係」，這兩個命題中要麼一個真，要麼兩個真，但必有一真，所以，題幹所説「三個判斷中只有一個是真的」必在（2）和（3）之中，從而可推出（1）必為假，根據「主任懂日語」為假，可推出「主任不懂日語」是客觀事實，由此可推（2）必為真。如果（2）為真，根據題幹，那麼（3）就為假，如果「有人懂日語」為假，可推出命題的矛盾命題「所有人不懂日語」為真，從而推出「副主任不懂日語」。

15. C

解析：本題的關鍵詞是「不確定的，含糊的，前途未卜的情境，總是試圖加以防止。」C選項中李先生總是感到緊張，沒有加以防止，不符合定義，所以選擇C。

PART ONE
題庫練習

PART TWO
模擬試卷

PART THREE
考生急症室

16. B

解析：解題的關鍵在於「急速文化變遷」「廣泛的」。A選項不廣泛，C選項根本不是「文化吸收的過程」。D也不是「文化吸收的過程」，而且也不廣泛。只有B完全符合，所以選擇B選項。

17. B

解析：本題的關鍵詞是「他人在場」並「促進其活動完成，提高其活動效率」，B選項中李老伯是動作的對象，因此不算是「他人在場」，與定義不符。所以選擇B選項。

18. C

解析：解題的關鍵在於利用的是「閒地」，從事的是「商品生產」，A項不是「商品生產」，B、D不是「閒地」。所以選擇C。

19. D

解析：題幹的關鍵是「職業活動中」和「接觸有毒有害物質」。A不是「職業活動中」，B、C不是「接觸有毒有害物質」。所以選擇D。

20. D

解析：解題的關鍵在於「公司權益受到侵害」「公司不起訴」。A、B是股東有權益受到侵害。C公司董事會起訴了。所以穩定擇D。

21. C

解析：A、B和D三個選項中，都沒有體現同一行業中的多家企業，所以排除。所以選C。

22. C

解析：A項屬於不隨意想像，所以排除。在B、D選項中，並沒有創造出新形象，所以不屬於想像。所以選C。

23. B

解析：在B選項中，公司將廣告交由同一家廣告公司管理不屬於幫助企業獲取、發展和籌劃智力和勞力資本，與定義不符，所以選擇B。

24. D

解析：A、B、C三個選項都沒有對當事人造成實質性的傷害，所以排除。所以選D。

25. A

解析：在B選項中，某人每天提早到辦公室上班為既定事實，所以沒有用理想化的自我、不現實的樂觀和擴大的可控性作為緩衝器，故排除B選項。在C選項中也沒有用理想化的自我、不現實的樂觀和擴大的可控性作為緩衝器故排除，D選項某男終獲成功，也不是不現實得樂觀，所以排除，所以選A。

26. A

解析：乙說去試試，沒想到會引燃油桶，所以是屬於疏忽大意的過失。

27. B

解析：要想知道人們在樹林裡是否比行人路上更安全，不單要知道受傷的人數，還必須知道兩個地方受傷人數佔總人數的比例，所以答案為B。

28. A

解析：這則定義中的關鍵點是為本單位謀取利益、由負責人員決定並以單位名義實施的行為。B項、C項和D項都不是為本單位謀取利益的行為。所以答案是A。

29. D

解析：D選項針對自然科學所佔有的比例得出這類圖書翻譯要求高，局功的周期長的原因。而A，B，C無從推出，題幹並沒有涉及到：「學術價值高市場風險高」、「先進科技類書籍」和「外國文學古典名著」。

PART ONE
題庫練習

PART TWO
模擬試卷

PART THREE
考生急症室

30. D

解析：題目中「幾乎每4人中便有1人會在一生中某個階段出現精神或行為問題」，但這些人並不是同時發病，所以全球抑鬱症患者並沒有達到1/4，A錯誤。題目中指出在2050年，心臟病會位於全球疾病排行榜第一，但並不是現在，所以B錯誤。題目中並沒有出高發病率是由於不能及時就醫造成的，所以C錯誤。非洲2,600萬抑鬱症，有90%的患者及時就醫，大概10%的人，即260萬人能夠及時就醫，因此不超過300萬是正確的，所以選D。

31. D

解析：奧肯定律是美國經濟學家阿瑟 • 奧肯發現的周期波動中經濟增長失業率之間的經驗關係，D項也是一種經驗獲得。

32. D

解析：（5）否定了（4）的後件，根據充分條件假言命題的推理規則，否定後件就能否定前件，所以可推出楊小姐出席；同理，由（3）可推出，李小姐出席；（2）否定了（1）的前件，根據必要條件假言命題的推理規則，否定前件就能否定後件，可推出李小姐、楊小姐或朱小姐不出席。由李小姐和楊小姐都出席，可推出朱小姐不出席，即D正確。

33. D

解析：從「丙最矮」、「乙比甲高」、「日本人比丁高」和「法國人最高」可知，丙、甲和丁都不是法國人，因此乙是法國人；由「丙最矮」和「日本人比丁高」可知，丙和丁不是日本人，則甲是日本人；由「丙最矮」和「英國人比俄國人高」可知，丙不是英國人，則丙是俄國人，丁是英國人。選D。

34. A

解析：全心全意為人民服務並不是公務員在履行職責、執行國家公務員的過程中可以做出一定行為、要求他人作出某種行為或者抑制一定行為的許可和保障，而是公務員的義務之一。

（二）Verbal Reasoning

(English) (6 questions)

Directions:

In this test, each passage is followed by three statements (the questions). You have to assume what is stated in the passage is true and decide whether the statements are either:

True (Box A): The statement is already made or implied in the passage, or follows logically from the passage.

False (Box B): The statement contradicts what is said, implied by, or follows logically from the passage.

Can't tell (Box C): There is insufficient information in the passage to establish whether the statement is true or false.

Example:

Passage 1

Researchers have successfully mapped the genetic code of the mouse and have generously elected to make their findings available to scientists everywhere. A genetic model of the mouse is of benefit to researchers because it is believed that one hundred million years ago, both mice and humans shared a common rodent-like ancestor. This makes mice ideal substitutes for humans in genetic studies. There-fore, the mouse is one of the most useful animals for studying cancer and other diseases.

1. The research findings show that humans and mice once had a common ancestor.

Answer : B

Explanation: The first sentence in the passage states that successful map-ping of the genetic code of mouse is the research result. From the second sentence of the passage, it is known that the sharing of a common rodent-like ancestor between mice and humans is a general belief. It is therefore a "false" statement.

2. It is likely that these findings, due to a similar shared genetic code, will result in research that will lead to the eradication of cancer.

Answer : C

Explanation: There is insufficient information in the passage to establish how useful this research will ultimately be.

Exercise:

Assume the passage is true and decide whether the statements are either: (A) True, (B) False or (C) Can't tell

Passage 1 (Question 1 to 3)

In the past few years, the government has turned to shock tactics to fight the battle against smoking and the burden smokers pose to the country's economy-manifested in sick days and health issues. The latest anti-smoking campaign showed tumours growing from a cigarette. Needless to say, this advert was contentious. On the one hand, the advert has become a viral sensation since its launch and has attracted plenty of attention to the subject. The government estimates approximately 300,000 people will attempt to quit smoking because of the campaign. One the other hand, some argue that using a hard-hitting approach is not the best way because people become very afraid. If someone thinks they might have cancer or a symptom of cancer, they're likely to push it out of their mind. Health experts claim the new campaign is not effective in targeting all smokers and recommend using it alongside other measures.

1. The advertisement is not effective because it scares people.

2. The advert was controversial.

3. The anti-smoking ad is funded by the Department of Health.

Passage 2 (Question 4 to 6)

People who drink coffee appear to live longer, as drinking coffee is associated with lower risk of death due to heart disease, cancer, stroke, diabetes, and kidney disease. People who consume a cup of coffee a day are 12 per cent less likely to die, compared to those who do not drink coffee. This association is even stronger for those who drink two to three cups a day - 18 per cent reduced chance of death. Lower mortality is present regardless of whether people drink regular or decaffeinated coffee, suggesting the association is not tied to caffeine.

4. Drinking coffee lowers the risk of dying of cancer.

5. The more coffee people drink, the more they are linked with longer lives.

6. People who drink two or three cups of coffee a day have a 6 per cent higher chance of living longer compared to people who drink one cup of coffee a day.

Passage 3 (Question 7 to 16)

Figures released by The Society of Motor Manufacturers and Traders (SMMT) show that new car registrations in the UK have increased in the first quarter of 2013 to 605,198 compared to the first three months of 2012 with 563,556 new cars being registered.

The registration of diesel cars has increased from 283,872 to 292,403; this indicates a 3% year on year increase. Due to the increased demand for the new downsized turbo petrol-engine cars, there has been a notable increase in 2013 of 32,570 (12.1%) cars to 304,356, compared to the amount of petrol cars being registered in 2012 being 271,481.

March 2013 saw Ford's Fiesta top the sales chart with 22,748 examples of the super-mini being registered in Britain, contributing to the 34,309 of the Fiesta models sold so far.

Figures from SMMT March sales show that even though that the Nissan Qashqai has been on sale in the UK for six years it can still compete against the best with being ranked 6th in the list with 8,465 registered and the BMW 1-series is placed 8th most popular with 7001 cars being registered. A highly productive year for drivers and sales teams alike!

7. There has been a bigger rise in diesel cars being registered than petrol cars.

8. Car sales dropped in the first three months of 2013 by 7.4% compared to 2012.

9. Ford has topped the cars sales with Fiesta for 2 years.

10. The Nissan Qashqui achieved more sales than the BMW 1-Series.

11. Ford has achieved over 34,000 sales of the Mondeo model.

12. Electric cars have increased in sales by over 12% since the last quarter.

13. A downsized engine for petrol cars has helped the sales.

14. Vehicles that have been on sale for over 5 years are still competing against best-selling models.

15. Northern parts of the UK have been major contributors for new car registrations.

16. The top selling super-mini of the 1stquarter of 2013 is produced by Ford.

PART ONE
題庫練習
PART TWO
模擬試卷
PART THREE
考生急症室

Passage 4 (Question 17 to 18)

Pride and Prejudice is a novel of manners by Jane Austen, first published in 1813. The story follows the main character, Elizabeth Bennet, as she deals with issues of manners, upbringing, morality, education, and marriage in the society of the landed gentry of the British Regency. Elizabeth is the second of five daughters of a country gentleman, Mr. Bennet, living in Longbourn. Set in England in the late 18th century, Pride and Prejudice tells the story of Mr. and Mrs. Bennet's five unmarried daughters after two gentlemen have moved into their neighbourhood. While Bingley takes an immediate liking to the eldest Bennet daughter, Jane, Darcy has difficulty adapting to local society and repeatedly clashes with the second-eldest daughter, Elizabeth.Pride and Prejudice retains a fascination for modern readers, continuing near the top of lists of "most loved books". It has become one of the most popular novels in English literature, selling over 20 million copies, and receiving considerable attention from literary scholars. Modern interest in the book has resulted in a number of dramatic adaptations and an abundance of novels and stories imitating Austen's memorable characters or themes.

17. Mr. Darcy eventually marries Elizabeth Bennet.

18. Jane Bennet is the eldest sister of Elizabeth.

Passage 5 (Question 19)

The Eiffel Tower is a wrought-iron lattice tower on the Champ de Mars in Paris, France. It is named after the engineer Gustave Eiffel, whose company designed and built the tower. Constructed in 1889 as the entrance to the 1889 World's Fair, it was initially criticised by some of France's leading artists and intellectuals for its design but has become a global cultural icon of Paris and one of the most recognizable structures in the world. The tower is the tallest structure in Paris, and the most-visited paid monument in the world: 6.98 million people ascended it in 2011. The tower received its 250 millionthe visitor in 2010. The tower is 324 meters tall, about the same height as an 81-storey building. Its base is square, 125 meters on a sid E. During its construction, the Eiffel Tower surpassed the Washington Monument to become the tallest human-made structure in the World, a title it held for 41 years until the Chrysler Building in New York City was built in 1930. Due to the addition of the aerial at the top of the tower in 1957, it is now taller than the Chrysler Building by 5.2 meters. Not including the broadcast antennae; it is the second-tallest structure in France, after the Millau Viaduct. The tower has three levels for visitors, with restaurants on the first and second levels. The top level's upper platform is 276 m above the ground, the highest accessible to the public in the European Union. Tickets can be pur-

PART ONE
題庫練習

PART TWO
模擬試卷

PART THREE
考生急症室

chased to ascend by stairs or lift (elevator) to the first and second levels. The climb from ground level to the first level is over 300 steps, as is the climb from the first level to the second. Although there is a staircase to the top level, it is usually accessible only by lift.

19. It takes over 600 steps to get to the second level of the Tower.

Passage 6 (Question 20-23)

The gas provider Centrica has seen its half-year profits fall by 20% to £992 million. Despite this it has decided to increase its dividend to shareholders to 3.9p per share. The rise in energy prices released before the announcement of Centrica's profits was designed, said the company, to restore "reasonable profitability". The chief executive, Sam Laidlaw, said, "We produced a good set of results in tough market conditions... we will continue to concentrate on improving customer service in British Gas..." This is an interesting statement given the record rise in prices to households. The announcement of profits which are not far off one billion pounds, a day after the announcement of the rise in price and the increase in dividend to shareholders, is likely to provoke angry reaction. Many will question the extent of the market power that Centrica possesses. There will also be those that will question what the phrase 'reasonable profitability' means in the context of profits of nearly one billion pounds.

20. Companies increase shareholder dividends when profits fall.

21. One way companies can restore profitability is to increase prices.

22. Centrica's half-year profits are not far off one million pounds.

23. Announcing large profits, soon after announcing price rises, is likely to provoke angry reaction.

Passage 7 (Question 24-25)

In December 2006, Mars were fined by the Environment Agency for breaches of European carbon trading rules. The agency failed to submit the correct permits for the carbon it emitted in 2005, the first year of the scheme. Mars were fined Euro 78,000 for failing to obtain allowances for almost 2,000 tonnes of carbon. Under the European Union (EU) system, firms are given allowances for the amount of carbon they can emit. If they exceed their allowance, they must buy additional permits from other companies to cover the shortfall. According to the Environment Agency, Mars produced 1,952 tonnes

PART ONE
題庫練習

PART TWO
模擬試卷

PART THREE
考生急症室

of carbon. It did not submit permits to cover these emissions until nearly eight months after the deadline. Mars said in their defence that they had "not been discharging excess greenhouse gases into the atmosphere, and in fact its emissions at the Peterborough facility have reduced."

24. The Environment Agency enforce European carbon trading rules.

25. In 2005, Mars produced 2,000 tonnes of carbon.

Answers :

1. C (Can't tell)

Explanation: The passage presents two opposite approaches of pros and cons to the anti-smoking campaign. The opinions presented are not addressed as absolute truths; therefore, it cannot be determined whether the advertisement is effective. The passage states: 'Some argue that using a hard-hitting approach is not the best way because people become very afraid'. The opinion of 'some' cannot be used as a generalisation. Thus, you Cannot Say whether the argument is True or False.

2. A (True)

Explanation: After presenting the advert, the passage states 'this advert was contentious', and follows this statement with a presentation of the attention the advert got and with the different opinions that surrounded it. Therefore, the advert was controversial, a synonym of contentious, and the argument is True.

3. C (Can't tell)

Explanation: The passage states: 'The government has turned to shock tactics to fight the battle against smoking and the burden smokers pose to the country's economy'. The passage does not state which department or ministry of the government is leading the campaign and funding it. Therefore, you Cannot Say whether the argument is True or False.

4. C (Can't tell)

Explanation: The passage states: 'People who drink coffee appear to live longer, as drinking coffee is associated with lower risk of death due to cancer'. In this quote and in the rest of the passage, drinking coffee is linked with lower cancer-related mortality, but it is not stated that drinking coffee will in fact lower the risk of dying of cancer. The passage states that there is a statistical connection, not a cause-and-effect one. People who drink coffee tend to be healthier, but you Cannot Say for sure that the reason for this is drinking coffee, as it could be because of something else.

5. C (Can't tell)

Explanation: The passage does state that people who drink one cup of coffee a day are 12 per cent less likely to die, while people who drink two to three cups a day have an 18 per cent reduced chance of death. However, the passage does not mention the chances of dying for people who drink more than three cupsof coffee every day—they could live lonfer and they could have a higher rate of mortality. You Cannot Say whether the argument is True or False based on the information in the passage.

6. B (False)

Explanation: The passage does state that people who drink one cup of coffee a day are 12 per cent less likely to die, while people who drink two to three cups a day have an 18 per cent reduced chance of death, and it seems that the difference is 18 – 12 = 6 per cent. However, it is most important that these chances mentioned in the passage are 'compared to those who do not drink coffee'. Thus, the difference between people who drink one cup of coffee a day, and people who drink two to three cups a day, is 6 per cent only when comparing them both to people who do not drink coffee. When comparing these two groups to one another, you get a different result.

7. B (False)

Explanation: Given information clearly communicates registration of diesel cars has shown a 3% year on year increase, as compared to a 12.1% increase for petrol cars. Also it is given that there has been an increase of 32,570 petrol car registration. From available data, the corresponding net increase in diesel car registration is 292,403 283,872= 8,531. So, rise is registration is less for diesel cars both in terms of absolute numbers and percentage. Hence the answer.

8. B (False)

Explanation: Statement shows an increase in the first three months of 2013 vs 2012.

9. C (Can't tell)

Explanation: Cannot say without further information. Statement makes no reference to '2 years'.

10. A (True)

Explanation: Statement quotes higher figure for BMW 1-Series.

11. C (Can't tell)

Explanation: Cannot say without further information. Statement makes no reference to Mondeo.

12. C (Can't tell)

Explanation: Cannot say without further information. Statement makes no reference to this.

13. A (True)

Explanation: Statement indicates this is true.

14. A (True)

Explanation: Statement supports this.

15. C (Can't tell)

Explanation: Cannot say without further information. Statement does not refer to 'North'.

16. A (True)

Explanation: Statement supports this.

17. C (Can't tell)

Explanation: Even though most readers of Pride and Prejudice are aware that Darcy marries Elizabeth, no information can be found about this event in the text. Therefore the answer is Cannot Say.

18. A (True)

Explanation: Bingley takes on the liking of the eldest daughter, Jane. Therefore we can conclude that the eldest sister of Elizabeth is indeed Jane.

PART **ONE**
題庫練習

PART **TWO**
模擬試卷

PART **THREE**
考生急症室

19. A (True)

Explanation: The climb from ground level to the first level is 'over 300 steps', as is the climb from the first to the secon D. Therefore, the answer is True.

20. B (False)

Explanation: According to the Environment Agency, Mars produced 1,952 tonnes of carbon.

21. A (True)

Explanation: The text states that rises in energy prices leads to increased profits.

22. B (False)

Explanation: The text states that Centrica's profits are not far off one billion pounds.

23. A (True)

Explanation: Refer to the statement: "The announcement of profits which are not far off one billion pounds, a day after the announcement of the rise in price and the increase in dividend to shareholders, is likely to provoke angry reaction."

24. A (True)

Explanation: The text states that the agency fined Mars for breaches of European carbon trading

25. B (False)

Explanation: According to the Environment Agency, Mars produced 1,952 tonnes of carbon.

(三)Data Sufficiency Test

PART ONE
題庫練習
PART TWO
模擬試卷
PART THREE
考生急症室

Directions :

In this test, you are required to choose a combination of clues to solve a problem.

Example :

John, Jack, Joseph and Jordan participated in a running race. Only the winner finished the race in 10 minutes. Which TWO pieces of information can tell you who the winner was?

(1) John finished the race in 16 minutes.

(2) Joseph finished the race earlier than Jack by 2 minutes.

(3) Jack finished the race later than Jordan by 4 minutes.

(4) Jordan finished the race earlier than John by 6 minutes.

(5) John finished the race later than Joseph by 4 minutes.

    A. (1) and (2)
    B. (1) and (4)
    C. (2) and (3)
    D. (2) and (5)
    E. (3) and (4)

Answer : B

Explanation: The winner finished the race in 10 minutes. Statements (1) and (4) will tell you that Jordan can finish the race in 10 minutes. The correct answer is therefore B.

Exercise:

In this test, you are required to choose a combination of clues to solve a problem.

1. **Mrs. Brown is dividing 50 students into 3 groups for a class project. How many children are in the largest group?**

   (1) The total number of children in the two smaller groups is equal to the number of children in the largest group.

   (2) The smallest group contains 6 children.

   A. Statement (1) ALONE is sufficient, but statement (2) alone is not sufficient.

   B. Statement (2) ALONE is sufficient, but statement (1) alone is not sufficient.

   C. BOTH statements TOGETHER are sufficient, but NEITHER statement ALONE is sufficient.

   D. EACH statement ALONE is sufficient.

   E. Statements (1) and (2) TOGETHER are NOT sufficient

2. **The total cost of food for the raccoons at the Altadena Wildlife Rescue has increased as the number of raccoons at the Rescue has increased. If it costs the same amount to feed each raccoon, is the cost of food for 7 raccoons more than $2,000 annually?**

   (1) It costs more than $1,000 annually to feed 4 raccoons.

   (2) It costs more than $1,500 annually to feed 5 raccoons.

A. Statement (1) ALONE is sufficient, but statement (2) alone is not sufficient.

B. Statement (2) ALONE is sufficient, but statement (1) alone is not sufficient.

C. BOTH statements TOGETHER are sufficient, but NEITHER statement ALONE is sufficient.

D. EACH statement ALONE is sufficient.

E. Statements (1) and (2) TOGETHER are NOT sufficient

3. Helena invested $8,000 in the Tallahassee City Bank at $z\%$ simple annual interest for one year with a yield of $450. How much should she invest at $s\%$ simple annual interest for one year to yield the same amount?

(1) $s/100 = 3/4$

(2) $s = .4z$

A. Statement (1) ALONE is sufficient, but statement (2) alone is not sufficient.

B. Statement (2) ALONE is sufficient, but statement (1) alone is not sufficient.

C. BOTH statements TOGETHER are sufficient, but NEITHER statement ALONE is sufficient.

D. EACH statement ALONE is sufficient.

E. Statements (1) and (2) TOGETHER are NOT sufficient

4. A certain voting bloc has how many voters?

(1) If no additional voters are added to the bloc, and 4 of the current voters leave the bloc, there will be fewer than 20 voters.

(2) If 4 more voters join the bloc and all of the present voters remain, there will be at least 27 voters.

A. Statement (1) ALONE is sufficient, but statement (2) alone is not sufficient.

B. Statement (2) ALONE is sufficient, but statement (1) alone is not sufficient.

C. BOTH statements TOGETHER are sufficient, but NEITHER statement ALONE is sufficient.

D. EACH statement ALONE is sufficient.

E. Statements (1) and (2) TOGETHER are NOT sufficient

5. Is y an integer?

(1) 7y is an integer.

(2) y/7 is an integer.

A. Statement (1) ALONE is sufficient, but statement (2) alone is not sufficient.

B. Statement (2) ALONE is sufficient, but statement (1) alone is not sufficient.

C. BOTH statements TOGETHER are sufficient, but NEITHER statement ALONE is sufficient.

D. EACH statement ALONE is sufficient.

E. Statements (1) and (2) TOGETHER are NOT sufficient.

6. **What is the value of positive two-digit integer x?**

(1) The sum of the two digits is 5.

(2) x is prime.

A. Statement (1) ALONE is sufficient, but statement (2) alone is not sufficient.
B. Statement (2) ALONE is sufficient, but statement (1) alone is not sufficient.
C. BOTH statements TOGETHER are sufficient, but NEITHER statement ALONE is sufficient.
D. EACH statement ALONE is sufficient.
E. Statements (1) and (2) TOGETHER are NOT sufficient.

7. Esther is giving Christmas presents to her family members. Each family member gets the same number of presents and no presents were leftover. If each family member gets at least one present, did each family member receive more than one present?

(1) Esther has forty Christmas presents to give out.

(2) If the number of family members were doubled, it would not be possible for each family member to get at least one present.

A. Statement (1) ALONE is sufficient, but statement (2) alone is not sufficient.

B. Statement (2) ALONE is sufficient, but statement (1) alone is not sufficient.

C. BOTH statements TOGETHER are sufficient, but NEITHER statement ALONE is sufficient.

D. EACH statement ALONE is sufficient.

E. Statements (1) and (2) TOGETHER are NOT sufficient.

8. A codebreaking device is made up of a rectangular box filled with x cylinders of ball bearings placed together such that the diameter of the bearings and the cylinders are equal, and the cylinders line up evenly, touching, with no extra room inside the device. If the cylinders are the same height as the box, and the box is 18 inches long and 10 inches wide, what's the value of x?

(1) 9 cylinders can line up along the length of the box.

(2) Each ball bearing has a radius of 1.

A. Statement (1) ALONE is sufficient, but statement (2) alone is not sufficient.

B. Statement (2) ALONE is sufficient, but statement (1) alone is not sufficient.

C. BOTH statements TOGETHER are sufficient, but NEITHER statement ALONE is sufficient.

D. EACH statement ALONE is sufficient.

E. Statements (1) and (2) TOGETHER are NOT sufficient.

9. **How many girls are members of both the Diving Team and the Swim Team?**

   (1) At a joint meeting of the Diving and Swim Teams, no members were absent and 18 girls were present.

   (2) The Diving Team has 27 members, one-third of whom are girls, and the Swim Team has 24 members, half of whom are girls.

   A. Statement (1) ALONE is sufficient, but statement (2) alone is not sufficient.
   B. Statement (2) ALONE is sufficient, but statement (1) alone is not sufficient.
   C. BOTH statements TOGETHER are sufficient, but NEITHER statement ALONE is sufficient.
   D. EACH statement ALONE is sufficient.
   E. Statements (1) and (2) TOGETHER are NOT sufficient.

10. **J and K are positive numbers. Is J/K>1?**

    (1) JK<1

    (2) J-K>0

    A. Statement (1) ALONE is sufficient, but statement (2) alone is not sufficient.
    B. Statement (2) ALONE is sufficient, but statement (1) alone is not sufficient.
    C. BOTH statements TOGETHER are sufficient, but NEITHER statement ALONE is sufficient.

D. EACH statement ALONE is sufficient.

E. Statements (1) and (2) TOGETHER are NOT sufficient.

11. A new Coffee Bean & Tea Leaf coffee drink consists only of certain amounts of espresso and sugar. What is the ratio of espresso to sugar in the new drink?

(1) There are 15 ounces of sugar in 35 ounces of the new drink.

(2) There are 40 ounces of espresso in 70 ounces of the new drink.

A. Statement (1) ALONE is sufficient, but statement (2) alone is not sufficient.

B. Statement (2) ALONE is sufficient, but statement (1) alone is not sufficient.

C. BOTH statements TOGETHER are sufficient, but NEITHER statement ALONE is sufficient.

D. EACH statement ALONE is sufficient.

E. Statements (1) and (2) TOGETHER are NOT sufficient.

12. How many in a group are women with blue eyes?

(1) Of the women in the group, 5 percent have blue eyes.

(2) Of the men in the group, 10 percent have dark-colored eyes.

A. Statement (1) ALONE is sufficient, but statement (2) alone is not sufficient.

B. Statement (2) ALONE is sufficient, but statement (1) alone is not sufficient.

C. BOTH statements TOGETHER are sufficient, but NEITHER statement ALONE is sufficient.

D. EACH statement ALONE is sufficient.

E. Statements (1) and (2) TOGETHER are NOT sufficient.

13. On a soccer team, one team member is selected at random to be the goalie. What is the probability that a substitute player will be the goalie?

(1) One-sixth of the team members are substitute players.

(2) 18 of the team members are not substitute players.

A. Statement (1) ALONE is sufficient, but statement (2) alone is not sufficient.

B. Statement (2) ALONE is sufficient, but statement (1) alone is not sufficient.

C. BOTH statements TOGETHER are sufficient, but NEITHER statement ALONE is sufficient.

D. EACH statement ALONE is sufficient.

E. Statements (1) and (2) TOGETHER are NOT sufficient.

14. A designer purchased 20 mannequins that each cost an equal amount and then sold each one at a constant price. What was the designer's gross profit on the sale of the 20 mannequins?

(1) If the selling price per mannequin had been double what it was, the gross profit on the total would have been $2400.

(2) If the selling price per mannequin had been $2 more, the store's gross profit on the total would have been $440.

A. Statement (1) ALONE is sufficient, but statement (2) alone is not sufficient.

B. Statement (2) ALONE is sufficient, but statement (1) alone is not sufficient.

C. BOTH statements TOGETHER are sufficient, but NEITHER statement ALONE is sufficient.

D. EACH statement ALONE is sufficient.

E. Statements (1) and (2) TOGETHER are NOT sufficient.

15. A shopping center increased its revenues by 10% between 2010 and 2011. The shopping center's costs increased by 8% during the same period. What is the firm's percent increase in profits over this period, if profits are defined as revenues minus costs?

(1) The firm's initial profit is $200,000.

(2) The firm's initial revenues are 1.5 times its initial costs.

A. Statement (1) ALONE is sufficient, but statement (2) alone is not sufficient.

B. Statement (2) ALONE is sufficient, but statement (1) alone is not sufficient.

C. BOTH statements TOGETHER are sufficient, but NEITHER statement ALONE is sufficient.

D. EACH statement ALONE is sufficient.

E. Statements (1) and (2) TOGETHER are NOT sufficient.

16. A certain number is not an integer. Is the number less than .4?

(1) The number rounded to the nearest tenth is .4.

(2) The number rounded to the nearest integer is 0.

A. Statement (1) ALONE is sufficient, but statement (2) alone is not sufficient.

B. Statement (2) ALONE is sufficient, but statement (1) alone is not sufficient.

PART ONE
題庫練習
PART TWO
模擬試卷
PART THREE
考生急症室

C. BOTH statements TOGETHER are sufficient, but NEITHER statement ALONE is sufficient.

D. EACH statement ALONE is sufficient.

E. Statements (1) and (2) TOGETHER are NOT sufficient.

17. **n is a positive number; z-15 is also a positive number; is z/n less than one?**

(1) **z-n>0**

(2) **n<15**

A. Statement (1) ALONE is sufficient, but statement (2) alone is not sufficient.

B. Statement (2) ALONE is sufficient, but statement (1) alone is not sufficient.

C. BOTH statements TOGETHER are sufficient, but NEITHER statement ALONE is sufficient.

D. EACH statement ALONE is sufficient.

E. Statement (1) and (2) TOGETHER are NOT sufficient to answer the question asked, and additional data are needed.

18. How many integers are there between m and n, exclusive, if m and n are themselves integers?

(1) m-n= 8

(2) There are 5 integers between, but not including, m-1 and n-1.

A. Statement (1) ALONE is sufficient, but statement (2) alone is not sufficient.

B. Statement (2) ALONE is sufficient, but statement (1) alone is not sufficient.

C. BOTH statements TOGETHER are sufficient, but NEITHER statement ALONE is sufficient.

D. EACH statement ALONE is sufficient.

E. Statements (1) and (2) TOGETHER are NOT sufficient.

19. For integers w, x, y, and z, is wxyz = -1?

(1) wx/yz=-1

(2) w=-1/x and y=1/z

A. Statement (1) ALONE is sufficient, but statement (2) alone is not sufficient.

B. Statement (2) ALONE is sufficient, but statement (1) alone is not sufficient.

C. BOTH statements TOGETHER are sufficient, but NEITHER statement ALONE is sufficient.

D. EACH statement ALONE is sufficient.

E. Statements (1) and (2) TOGETHER are NOT sufficient.

PART ONE
題庫練習
PART TWO
模擬試卷
PART THREE
考生急症室

20. If the product of j and k does not equal zero, is j<0 and k>0?

    (1) (-j, k) lies above the x-axis and to the right of the y-axis.

    (2) (j, -k) lies below the x-axis and to the left of the y-axis.

    A. Statement (1) ALONE is sufficient, but statement (2) alone is not sufficient.
    B. Statement (2) ALONE is sufficient, but statement (1) alone is not sufficient.
    C. BOTH statements TOGETHER are sufficient, but NEITHER statement ALONE is sufficient.
    D. EACH statement ALONE is sufficient.
    E. Statements (1) and (2) TOGETHER are NOT sufficient.

21. What is the value of xn-ny-nz?

    (1) x-y-z=10

    (2) n=5

    A. Statement (1) ALONE is sufficient, but statement (2) alone is not sufficient.
    B. Statement (2) ALONE is sufficient, but statement (1) alone is not sufficient.
    C. BOTH statements TOGETHER are sufficient, but NEITHER statement ALONE is sufficient.
    D. EACH statement ALONE is sufficient.
    E. Statement (1) and (2) TOGETHER are NOT sufficient to answer the question asked, and additional data are needed.

22. If A and B are integers, is B>A?

(1) B>10

(2) A<10

A. Statement (1) ALONE is sufficient, but statement (2) alone is not sufficient.

B. Statement (2) ALONE is sufficient, but statement (1) alone is not sufficient.

C. BOTH statements TOGETHER are sufficient, but NEITHER statement ALONE is sufficient.

D. EACH statement ALONE is sufficient.

E. Statement (1) and (2) TOGETHER are NOT sufficient to answer the question asked, and additional data are needed.

23. Is (-x) a negative number?

(1) 4x(2)-8x>(2x)(2)-7x

(2) x+2>0

A. Statement (1) ALONE is sufficient, but statement (2) alone is not sufficient.

B. Statement (2) ALONE is sufficient, but statement (1) alone is not sufficient.

C. BOTH statements TOGETHER are sufficient, but NEITHER statement ALONE is sufficient.

D. EACH statement ALONE is sufficient.

E. Statement (1) and (2) TOGETHER are NOT sufficient to answer the question asked, and additional data are needed.

PART **ONE**
題庫練習

PART TWO
模擬試卷

PART THREE
考生急症室

Answers:

1. A

Explanation: The first statement allows us to express the total number of children in the first two groups in terms of the number of children in the third group. Let's call the smaller groups A and B, and the largest group C. Thus, we can express the first statement: A+B=C.

We know from the question-stem that A + B + C = 50, so we know C=50–(A+B).

Using substitution, C=50–C.

2C=50

C=25. Sufficient.

The second statement tells us the value of A, the smallest group.

This only tells us that 6+B+C=50. Without knowing B, we cannot determine a unique value for C.

2. B

Explanation: (1) If it costs more than $1,000 annually to feed 4 raccoons, we do not have enough information to answer either yes or no to the original question. It could cost $2,000 to feed 4 raccoons, in which case it WOULD cost more than $2,000 to feed 7 raccoons. Or, it could cost only $1,000 and one cent to feed 4 raccoons, in which case feeding 3 more would be less than an additional $1,000, and the answer would be no. This statement is insufficient.

(2) If it costs more than $1,500 annually to feed 5 raccoons, then the smallest cost for each animal is a little over $300. $300x7 raccoons = $2,100. Sufficient.

3. D

Explanation: The missing information here is the value of s. Both statements allow us to find s. Statement (1) allows us to do so by simplifying and solving the equation for s. From Statement (2) we know we can find "z" from the information given in the question stem. Once we find z, we can plug in for s. From the question stem we can calculate z% as follows:

Simple interest = principal x rate x time. $450=$8000x(z/100)x1, so z=5.625%. Now that we know z, we can plug in to solve for s in Statement (2).

4. C

Explanation: (1) If no additional voters are added to the bloc, and 4 of the current voters leave the bloc, there will be fewer than 20 voters.

We can translate the given information into an inequality: $x-4<20$, where "x" is the number of current voters. We know $x<24$, but we cannot determine an exact value for x.

(2) If 4 more voters join the bloc and all of the present voters remain, there will be at least 27 voters.

We can translate the given information into an inequality: $x+4\geq27$. "At least" means there could be 27 OR more than 27 in the bloc. This inequality simplifies to $x\geq23$. We do not know the exact value of x based on this inequality.

Combining both statements we know $23\leq x<24$. If x must be less than 24, but greater or equal to 23, the only number that satisfies both conditions is 23.

If you chose (D), keep in mind that each statement alone only allows us to limit the range of possible values for "x," but not find the actual numerical value. For a "value" DS question, if more than one number is possible, the statement cannot be sufficient.

If you chose (E), you may not have realized that we could have expressed the information in the statements as inequalities. Both statements combined then allow us to limit the range of possible values to one, so combined they are sufficient.

5. B

Explanation: (1) 7y is an integer.

If you chose (A), note that it is possible for 7y to be an integer when y is an integer. For example, if y=1, 7y=7. However, it is also possible for 7y to be an integer when y is not an integer. For example, if 7y=1, then y =1/7.

(2) y/7 is an integer.

If y/7 is an integer, then y must be a multiple of 7. All multiples of 7 are themselves integers (7, 14, 21, etc.).

If you chose (C), you failed to recognize that Statement (2) was sufficient by itself, as there is no value we can choose for y that makes y/7 an integer that is not itself an integer. Picking numbers can help you see this relationship more clearly.

6. E

Explanation: If you chose (A), from Statement 1, the possible values for the digits are 0 and 5, 1 and 4, or 2 and 3. So the possible numbers are: 50, 14, 41, 23, or 32. We don't know which of these is x.

If you chose (B), from Statement 2, we know x is prime, but there are many two-digit prime numbers: 11, 13, 17, 19, etc.

If you chose (C), remember that both 23 and 41 are prime numbers.

If you chose (D), both statements offer limiting information, but because this is a "value" question, a statement can only be considered sufficient if it allows us to limit our range of possible x's to ONE value only.

The answer is (E). Statement 1 limits our possible x's to 5 integers, and Statement 2 narrows that list to 2 integers. However, we still do not know whether x is 23 or 41.

7. B

Explanation: We know from the question that each family member got at least 1 present, that they all got the same number of presents, and that no presents were left over. Based on this information, we can write the following inequality:

(number of presents / number of family members) ≥ 1

(1) The first statement would be useful if we needed to know the number of family members, but it doesn't help to answer this yes/no question.

(2) Based on the information in statement (2), we can write the following inequality:

[number of presents/(2 x number of family memebers)] < 1

OR

[number of presents/(number of family memebers)] < 2

Combining the information in statement (2) and the information in the question stem, we find that:

1 ≤ [number of presents/(number of family memebers)] < 2

Since each family member got at least one present, and the number of presents per family member is less than 2, we can conclude that each family member received only one present.

The answer to the original question is NO. However, if we can answer YES/NO to a data sufficiency question based on the information in a statement, then that statement is sufficient. Statement (2) is sufficient.

8. D

Explanation: Statement (1) is sufficient. If the box is 18 inches long and 9 cylinders fit along that length, then they must each have a diameter of 2.

Statement (2) is also sufficient. In order to find the value of x, we need to know the diameter of each cylinder. This is given by Statement (2), since twice the radius of the ball bearing will equal the cylinder's diameter. If each cylinder has a diameter of 2, then 9 will fit along the length of the box and 5 will fit along the width. A total of 9x5=45 cylinders will fit inside the box.

9. C

Explanation: To find the number of girls who are members of BOTH teams, we must find the overlap of this set. From Statement (1), we know that there are 18 girls total. Some are members of the Diving Team-only, some are members of the Swim Team-only, and some are members of both. But we do not have enough information to determine the number of girls who are members of both teams. Statement (1) alone is not sufficient.

From Statement (2), we know that 1/3 of the 27 Diving Team members are

PART ONE
題庫練習
PART TWO
模擬試卷
PART THREE
考生急症室

girls, so there are 9 girls in the Diving Team. We're also told that 1/2 of the 24 Swim Team members are girls, so there are 12 girls in the Swim Team. Combined, that is a head-count of 9+12=21 girls. However, we still cannot determine how many girls are members of both teams. Statement (2) alone is not sufficient.

Combining both statements, we know that there are only 18 girls total who are members of these two clubs. Therefore, the extra 3 girls from our "head-count" must come from the number of girls who are members of both clubs.

10. B

Explanation: Statement (1) says that JK<1. In order for this to be true, at least one of these numbers must be a positive fraction. We can quickly choose numbers to test this: If J=1 and K=1/2, their product is 1/2 and this less than 1. In this case, the answer to the question would be "YES" since 1÷1/2=2, which is greater than 1. But what if J=1/2 and K=1?

J/K would be equal to 1/2. In this case, the answer to the question would be "NO". Since the answer to the question can be "YES" or "NO" depending on the values of J and K, Statement (1) alone is not sufficient to answer the question.

Statement (2) tells us that J-K>0. Let's quickly choose values again. If J=1 and K=1/2, we satisfy the statement and get a "YES" answer. In fact, for all values we choose we will get a "YES" since J-K>0 can be manipulated to read J>K. If J is always greater than K, then J/K will always be greater than 1. Statement (2) Alone is sufficient to answer the question.

11. D

Explanation: We do not need to know the exact amount of espresso and sugar in the new drink. We only need to know the relationship between the amounts, since this question only asks for the ratio of espresso to sugar in the new drink.

If you chose (A), remember that we can find the ratio between ingredients as soon as we know the amount of each ingredient in the beverage or if we know the amount of one ingredient in relation to the set amount of the drink (part to part , or one part to whole). Since there are only two ingredients in the new drink, 20 ounces must be espresso. The ratio in (1) would be 20:15, or 4:3. However, (B) is also sufficient, since 30 ounces would be sugar.

If you chose (B), since the new drink consists only of espresso and sugar, we can find the ratio of espresso to sugar. The 30 ounces difference here must be made up of sugar. Therefore, the ratio is 40:30, or 4:3.

If you chose (C), you failed to recognize that each statement alone is sufficient. This is because the question is asking only for a ratio. We don't need all of the "real-world" values to come up with a ratio, only the part-to-part or part-to-whole in a given circumstance to express the ratio. Each statement is sufficient to do this.

If you chose (E), you missed the idea that you can figure out the amount of espresso and the amount of sugar for a given amount of the new drink based on either statement (you just need to subtract).

12. E

Explanation: There are three numbers we must know in order to find the percent of women with blue eyes: number of men in the group, number of women in the group, number of women with blue eyes.

If you chose (A), we know that 5% of the women have blue eyes, but we do not know how many members of the group are women, thus we cannot answer the question.

If you chose (B), this only gives us information on the men in the group, but the question concerns the number of women who fit a certain criteria.

If you chose (C), the information in the second statement does not tell us anything about the women in the group, and the first statement only tells us the percentage with blue eyes, not enough to determine the actual number.

If you chose (D), both statements are insufficient for different reasons. Statement (1) tells us the percentage, but not the actual number of women in the group, so we cannot turn that percentage into a number as this "value" question requires. Statement (2) does not relate at all to the women in the group and is therefore insufficient.

PART ONE
題庫練習
PART TWO
模擬試卷
PART THREE
考生急症室

13. A

Explanation: The probability that a substitute will be chosen can be found if we know the ratio of substitutes to total members OR if we know both values exactly. Here we are given the ratio=1/6, so it is sufficient. If you chose (B), we know 18 are NOT substitutes, but we do not know how many ARE substitutes, so we cannot determine the probability.

14. B

Explanation: This question asks about gross profit, which we know is derived from subtracting the total cost from the total selling price. If we know the cost of each mannequin and the selling price of each mannequin we can determine the designer's gross profit. From the given information, we can write the following equation: $P=20(s–c)$, where s=selling price and c=cost. So either we'll need a value for s and a value for c, or we'll need the value of $(s–c)$.

Statement (1) tells us that $2400=(20(2s–c))$ or $2400=40s–20C$. We can divide both sides by 20 and simplify the equation to get: $120=2s–C$. We still don't know s and C. Insufficient.

Statement (2) tells us that $440=20(s+2–c)$. Let's simplify:

$440=20s+40-20c$

$400=20s-20c$

$400=20(s-c)$

Sufficient. Even though we didn't solve for s and c separately, we were able to find the value of $(s-c)$.

15. B

Explanation: Let' start with our most basic Profit formula: Profit=Revenue-Cost.

Using Statement (1), we can say that $200,000=R- C$.

Profit in 2010=R-C

Profit in 2011=1.1R-1.08C

Percentage change=[(Final-Original) / Original]x100%

Percentage change=[(1.1R-1.08C)-(R-C) / R-C]x100%

Percentage change=[(0.1R-0.08C)/(R-C)]x100%

Percentage change=[(0.1R-0.08C)/200,000]x100%

Without knowing R or C, we cannot determine the percentage change in this case. Notice that if we knew R or some other relationship between R and C, we could substitute into the equation R-C=200,000 to solve for the missing piece.

Statement 1 is insufficient.

Using Statement (2), R=1.5C

Percentage change=[(Final-Original)/Original]x100%

Percentage change=[[(1.1R-1.08C)-(R-C)]/R-C]x100%

Percentage change=[(0.1R-0.08C)/(R-C)]x100%

Percentage change=[(0.1(1.5C)-0.08C) / (1.5C-C)]x100%

Percentage change=(0.07C/0.5C)X100%=14%

The last step was not necessary if you realized that substituting R=1.5C into the expression for calculating the percentage change will give you an expression in which the Cs cancel out, giving you an actual percentage.

Statement 2 is sufficient.

16. E

Explanation: If you chose (A), we know the "nearest" tenth is .4, so this means the number is between .35 and .45. If it were smaller than .35 (for example, .34), we would round down to .3, and if it were greater than .45 (for example, .47), we would round up to .5. Since .4 is between .35 and .45, we can't tell if the value of the number is less than or greater than (or equal to) .4.

If you chose (B), we know the "nearest" integer is 0, so the number must be between -0.5 and .5. While most of the values in this range are less than .4, the values between .4 and .5 are greater than .4.

If you chose (C), remember we can express the information from the first statement as an inequality: .35≤x<.45. And we can express the information from the second statement as an inequality: -0.5<x<.5. Since the range of values for the first inequality is within the range of the second inequality, the range of values that satisfy both inequalities will be the same as the range of values in the first inequality (.35≤x<.45). Since we already determined that the first inequality is not sufficient, combining both inequalities will not be sufficient.

If you chose (D), you may have missed the possible values between .4 and .45 in Statement 1, and the range of possible values between .4 and .5 in Statement 2.

(E) is the correct response. Even combined, we get a range between .35 and .45. There are values that are possible both above and below (and equal to) .4.

17. D

Explanation: 1. The value of a fraction is less than one if its numerator is smaller than its denominator. For example, 4/6 is less than one because 4<6. So, the question at hand can be simplified to: is $z<n$?

2. Evaluate Statement (1) alone:

(a) Statement (1) can be re-arranged:

$z-n>0$

$z>n$

(b) Since $z>n$, you can definitively answer no to the question: "is $z<n$?"

(c) Statement (1) is SUFFICIENT.

3. Evaluate Statement (2) alone:

(a) Based upon the question, since z-15 is a positive number, the following inequality must hold:

$z-15>0$

$z>15$

(b) Statement (2) says:

$n<15$

(c) Since $z>15$ and $n<15$, you know that $z>n$

(d) You can definitively answer no to the question: "is $z<n$?"

(e) Statement (2) is SUFFICIENT.

4. Since Statement (1) alone is SUFFICIENT and Statement (2) alone is SUFFICIENT, answer D is correct.

18. D

Explanation: This question is provides a great opportunity to try out numbers.

Statement 1: Let's say m=9 and n=1, then m-n=9-1=8. There are 7 integers between 1 and 9. If we choose a different set of integers: let's say m=-2 and n=-10. There are still 7 integers between these two numbers.

Statement 2: If there are 7 integers between m-1 and n-1, then there will still be 7 integers between m and n. This is sufficient for the same reason that statement 1 was sufficient.

19. B

Explanation: We can plug in these values for "w" and "y" in the original equation:

(-1/x)(x)(1/z)(z)=-1. The x's and z's cancel out so that we get: (-1)(1)=-1. The answer will always be "Yes" no matter what the actual values are for w, x, y, and z. This is sufficient.

If you chose (A), remember this is a Y/N question. The stem does not tell us any information about what numbers w, x, y, and z are. This is a great question to pick numbers! As long as we can choose two sets of numbers: one that gives us an outcomes of "Yes" and the other that gives an outcome of "No," we know the statement is insufficient.

Let's choose w=1, x=1, y=1, and z=-1. These numbers satisfy the condition in Statement (1) and allow us the answer the question "Yes." However, if we chose w=2, x=2, y=2, and z=-2, we would answer the question "No." Therefore, this Statement must be insufficient.

If you chose (C), you may not have realized that we could substitute the value in Statement (2) to simplify wxyz=-1. By doing this, we realize that for ALL integers, b and d will cancel out leaving us with (-1)(1)=-1, which is always true.

If you chose (D), you correctly saw that substitution allows Statement (2) to be sufficient, but Statement (1) is not sufficient. Depending on what integers we select for the variables we can make their product equal or not equal to -1.

If you chose (E), you missed that Statement (2) is sufficient once we substi-

PART **ONE**
題庫練習

PART **TWO**
模擬試卷

PART **THREE**
考生急症室

tute it into the original equation. Sometimes statements will be unexpectedly sufficient in this way on the GMAT, even though we don't know the actual values for the variables!

20. D

Explanation: Each statement alone is sufficient.

Statement 1: If (-j, k) lie above the x-axis and to the right of the y-axis (that is, in the first quadrant), then we can write the following two inequalities: -j>0, and k>0 OR j<0 and k>0. Statement 1 is sufficient to answer the question.

Statement (2): If (j, -k) is in the third quadrant (as the information in statement 2 implies), then j = negative, and –k = negative.

We can write the following inequalities: j<0, and –k<0, OR j<0, and k>0. Statement 2 alone is sufficient to answer the question.

21. C

Explanation:

1. Factor the original equation: xn-ny-nz=n(x-y-z)

2. If we know the value of both n and x-y-z, we can determine the value of xn-ny-nz

3. Evaluate Statement (1) alone:

(a) Since x-y-z=10, based upon the above factoring:

xn-ny-nz=n(10)

However, we do not know the value of n so we cannot solve for the value of xn-ny-nz.

(b) Statement (1) is NOT SUFFICIENT.

4. Evaluate Statement (2) alone:

(a) Since n = 5, based upon the above factoring:

xn-ny-nz=5(x-y-z)

However, we do not know the value of x-y-z so we cannot solve for the value of xn-ny-nz.

(b) Statement (2) is NOT SUFFICIENT.

5. Evaluate Statements (1) and (2) together:

(a) Since n=5 and x-y-z=10, based upon the above factoring: xn-ny-nz=n(x-y-z)=5(10)=50

(b) Statements (1) and (2), when taken together, are SUFFICIENT.

6. Since Statement (1) alone is NOT SUFFICIENT and Statement (2) alone is NOT SUFFICIENT, but Statements (1) and (2), when taken together, are SUFFICIENT, answer C is correct.

22. C

Explanation: 1. Evaluate Statement (1) alone.

(a) Statement (1) simply says that B>10. It provides no information about the value of A, making a comparison between B and A impossible.

(b) If B=12 and A=5, then the answer to the question "is B>A?" would be yes. However, if B=15 and A=20, then the answer to the question "is B>A?" would be no.

(c) Since different legitimate values of A and B produce different answers to the question, Statement (1) is NOT SUFFICIENT.

2. Evaluate Statement (2) alone.

(a) Statement (2) simply says that A<10. It provides no information about the value of B, making a comparison between B and A impossible.

(b) If B=12 and A=5, then the answer to the question "is B>A?" would be yes. However, if B=1 and A=9, then the answer to the question "is B>A?" would be no.

(c) Since different legitimate values of A and B produce different answers to the question, Statement (2) is NOT SUFFICIENT.

3. Evaluate Statements (1)and (2) together.

(a) When taking Statements (1) and (2) together, you know:

B>10 and A<10

(b) So, you know that B> A. Statements (1) and (2), when taken together, are SUFFICIENT.

4. Since Statement (1) alone is NOT SUFFICIENT and Statement (2) alone is NOT SUFFICIENT yet Statements (1) and (2), when taken together, are SUFFICIENT, answer C is correct.

PART ONE
題庫練習

PART TWO
模擬試卷

PART THREE
考生急症室

23. A

Explanation: 1. Evaluate Statement (1) alone:

(a) Simplify the inequality:

4x(2)-8x>(2x)(2)-7x

4x(2)-8x>4x(2)-7x

-8x>-7x

-8x+8x>-7x+8x

0>x

x<0

(b) Since X is less than zero, X is a negative number. This means that negative X is a positive number since multiplying a negative number by a negative number (i.e., -1) results in a positive number.

(c) Statement (1) alone is SUFFICIENT.

2. Evaluate Statement (2) alone:

(a) Simplify the inequality:

x+2>0

x>-2

(b) Since we cannot be sure whether X is negative (e.g., -1) or positive (e.g., 2), we cannot be sure whether negative X is positive or negative.

(c) Statement (2) alone is NOT SUFFICIENT.

3. Since Statement (1) alone is SUFFICIENT and Statement (2) alone is NOT SUFFICIENT, answer A is correct.

(四) Numerical Reasoning

PART **ONE**
題庫練習

PART *TWO*
模擬試卷

PART THREE
考生急症室

Directions:

Each question is a sequence of numbers with one or two numbers missing. You have to figure out the logical order of the sequence to find out the missing number(s).

Example:

2，6，10，?，18，22，?

    A. 12, 24

    B. 14, 26

    C. 16, 28

    D. 18, 30

    E. 20, 32

Answer: B

Explanation: It is a mathematical sequence with each number increasing by 4. The first missing number is 10 + 4 which is 14, and the second missing number is 22 + 4 which is 26. The answer is therefore B.

練習題：

(1) 1.01，1.02，2.03，3.05，5.08，(　)

    A. 8.13

    B. 8.013

    C. 7.12

    D. 7.012

(2) 28，16，12，4，8，(　)

    A. -8

    B. 6

    C. -4

    D. 2

(3) 29，21，15，(　)，9

    A. 17

    B. 11

    C. 25

    D. 7

(4) 3，8，22，62，178，(　)

    A. 518

    B. 516

    C. 548

    D. 546

(5) 0，2，2，5，4，7，(　)

    A. 6

    B. 5

    C. 4

    D. 3

(6) 2，7，19，60，176，(　)

    A. 530

    B. 531

    C. 532

    D. 533

(7) 1，-4，4，8，40，（ ）

    A. 160

    B. 240

    C. 320

    D. 480

(8) 2，2，7，9，16，20，（ ）

    A. 28

    B. 29

    C. 30

    D. 31

(9) 243，162，108，72，48，（ ）

    A. 26

    B. 28

    C. 30

    D. 32

PART ONE
題庫練習
PART TWO
模擬試卷
PART THREE
考生急症室

(10) 2，4，4，8，16，( )

A. 48

B. 64

C. 128

D. 256

(11) 1，1，6，21，81，306，( )

A. 1017

B. 1161

C. 1285

D. 1527

(12) 0，2，2，4，6，( )

A. 4

B. 6

C. 8

D. 10

(13) 2/3，4/7，6/11，8/15，( )

    A. 1/2

    B. 11/23

    C. 11/19

    D. 10/19

(14) 3，9，4，16，( )，25，6，( )

    A. 5，36

    B. 10，36

    C. 6，25

    D. 5，30

(15) 1，6，6，36，( )，7776

    A. 96

    B. 216

    C. 866

    D. 1776

PART ONE
題庫練習
PART TWO
模擬試卷
PART THREE
考生急症室

(16) 1，2，6，21，109，768，（　）

A. 8448

B. 8450

C. 8452

D. 8454

(17) 1，-3，3，3，9，（　）

A. 28

B. 36

C. 45

D. 52

(18) 2187，729，243，81，27，（　）

A. 3

B. 6

C. 9

D. 12

(19) 1/3，1/2，5/11，7/18，1/3，(　)

    A. 11/38

    B. 13/34

    C. 5/12

    D. 7/15

(20) 7，15，29，59，117，(　)

    A. 227

    B. 235

    C. 241

    D. 243

(21) 1.5，9.5，24.5，48.5，(　)

    A. 83.5

    B. 88.5

    C. 98.5

    D. 68.5

PART ONE
題庫練習

PART TWO
模擬試卷

PART THREE
考生急症室

(22) 2246，3164，5180，6215，（ ）

    A. 5711

    B. 7132

    C. 8591

    D. 9125

(23) 1，7，7，9，3，（ ）

    A. 7

    B. 11

    C. 6

    D. 1

(24) 1269，999，900，330，（ ）

    A. 190

    B. 270

    C. 299

    D. 1900

(25) 10，21，44，65，（　）

    A. 122

    B. 105

    C. 102

    D. 90

(26) 1，10，7，10，19，（　）

    A. 16

    B. 20

    C. 22

    D. 28

(27) 1，2，9，64，625，（　）

    A. 1728

    B. 3456

    C. 5184

    D. 7776

PART **ONE**
題庫練習

PART *TWO*
模擬試卷

PART THREE
考生急症室

(28) 6，14，22，( )，38，46

    A. 30

    B. 32

    C. 34

    D. 36

(29) 5，14，65/2，( )，217/2

    A. 62

    B. 63

    C. 64

    D. 65

(30) 124，3612，51020，( )

    A. 7084

    B. 71428

    C. 81632

    D. 91836

答案及解析：

1. A

解析：本題是小數數列。整數部分數列為：1、1、2、3、5、(8)，後一項等於相鄰前兩項之和；小數部分數列為：01、02、03、05、08、(13)，同樣為後一項等於相鄰前兩項之和，因此原數列的下一項為8.13，故正確答案為A。

2. C

解析：本數列是遞推數列。遞推規律為：前兩項之差等於下一項。具體規律為：28-16=12，16-12=4，12-4=8，則原數列下一項為4-8=-4，故正確答案為C。

3. B

解析：前一項減後一項分別得到8，6，4，2，括號內應填11。

4. A

解析：3x3-1=8，8x3-2=22，22x3-4=62，62x3-8=178，178x3-16=518，其中減數1、2、4、8、16是公比為2的等比數列。

5. A

解析：奇數項：0、2、4、(6)為等差數列；偶數項：2、5、7為簡單遞推數列，第三項等於前兩項之和。故正確答案為A。

6. D

解析：原數列有如下關係：7＝2x3＋1，19＝7x3-2，60＝19x3＋3，176=60x3-4，(533)=176x3+5。

7. C

解析：兩兩相除的結果為-4,-1,2,5,8，故答案為40x8=320。

PART ONE
題庫練習

PART TWO
模擬試卷

PART THREE
考生急症室

8. B

解析：原數列為二級做和數列。數列中相鄰兩項做和構成新數列：4、9、16、25、36、(49)，為平方數列，所以未知項為49-20=29。故正確答案為B。

9. D

解析：原數列為公比為2/3的等比數列，因此未知項為48x(2/3)=32，故正確答案為D。

10. B

解析：本題為遞推數列。遞推規律為：數列相鄰兩項乘積的一半得到數列的下一項。具體規律為：(2x4)/2=4，(4x4)/2=8，(4x8)/2=16，因此原數列的下一項為：(8x16)/2=64，故選擇B選項。

11. B

解析：從第三項開始，每一項等於前兩項之和的3倍，依此類推，所求項為(81+306)x3=1161，故答案是B。

12. D

解析：原數列為做和遞推數列：0+2=2，2+2=4，2+4=6，即從第三項起，每一項都等於它前兩項的和，故未知項為4+6=10，所以正確答案為D。

13. D

解析：這是一組分數數列，分子分母分別各構成一組新數列。分子數列為2、4、6、8、( )，是一組公差為2的等差數列，括號內當為10。分母數列為3、7、11、15、( )，是一組公差為4的等差數列，括號內當為19，可知正確答案為D項。

14. A

解析：奇數項：3、4、(5)、6為等差數列；偶數項：9、16、25、(36)為平方數列，因此原數列第5項為5，第8項為36。故正確答案為A。

15. B

解析：從前四個數字可以看出規律，即前兩項的乘積等於第三項，故 6x36=216，再用所給數列中的第六項來進行驗證，36x216=7776，也正確，故正確選項為B。

16. D

解析：原數列有如下關係：1x1+1＝2，2x2+2＝6，6x3+3＝21，21x5+4=109，109x7+5=768，故下一項為768x11+6，根據首尾數法可知，該值的尾數為8x1+6的尾數，即為4。

17. C

解析：原數列為二級做商數列。數列中相鄰兩項之間，後項除以前項構成新數列：-3、-1、1、3、(5)，為等差數列，所以未知項為9x5=45。故正確答案為C。

18. C

解析：很顯然整列數字都是3的三次方，依次為，故本題正確答案為C。

19. A

解析：將數列變為1/3、3/6、5/11、7/18、9/27，分子列為等差數列，分母列為二級等差數列，則未知項為11/38，因此正確答案為A。

20. B

解析：本題為2倍遞推數列，依次±1進行修正。具體規律為：7x2+1=15，15x2-1=29，29x2+1=59，59x2-1=117。因此，原數列下一項為：117x2+1=235，故正確答案為B。

21. A

解析：一次作差分別為8,15,24；8=2x4,15=3x5,24=4x6，所以下一個相差的數是5x7=35，所以結果為83.5。

PART ONE
題庫練習

PART TWO
模擬試卷

PART THREE
考生急症室

22. A

解析：此數列數位相同、大小變化紊亂，屬於組合拆分數列的特徵，嘗試拆分無明顯規律，只能是數位組合，數列中各個數的數位之和=14，只有A選項符合。因此，本題答案選擇A選項。

23. A

解析：觀察原數列可知，從第三項起，當前項是取前一項與更前一項乘積的個位數，即：1、7、(1x7=7，取個位數7)、(7x7=49，取個位數9)、(7x9=63，取個位數3)、( )。則空缺項應為9x3=27的個位數，即為7，正確答案為A。

24. B/D

解析：本題為遞推數列。遞推規律為：前兩項之差乘以10/3等於下一項。具體規律為：(1269-999)x10/3=900，(999-900)x10/3=330，因此原數列未知項為(900-330)x10/3=1900，故正確答案為D。另外觀察題幹，都是3的倍數，而選項中只有B是3的倍數，所以選擇B也可以。

25. C

解析：因式分解數列。列中的項一次拆分為2x5、3x7、4x11、5x13，我們發現2、3、4、5是等差數列，下一項為；5、7、11、13為質數列，下一項為17，所以答案為6x17=102，即C選項。

26. A

解析：

解析一：把本題當做二級循環數列，後項減去前項，得到9，-3,3,9，(-3)，為周期循環數列，因此所求項為19-9=16，答案為A。

解析二：把本題當做和數列。相鄰三項加和，可得18，27，36，(45)，為等差數列，從而推知答案為A選項。

27. D

解析：冪次數列。數列的每一項都是冪次數，分別為1的0次方，2的1次方，3的2次方，4的3次方，5的4次方，下一項的底數應該是1、2、3、4、5的後一個，即6，指數為0、1、2、3、4、的後一個，即5。所以答案為6的5次方，根據尾數法，從B、D中選，6的5次方為216x36大於6000，所以D為正確答案。

28. A

解析：原數列為等差數列，相差為8，所以未知項為22+8=38-8=30。故正確答案為A。

29. B

解析：5=10/2 ,14=28/2 , 65/2,（126/2), 217/2，分子－10=2^3+2；28=3^3+1；65=4^3+1；(126)=5^3+1；217=6^3+1；其中2、1、1、1、1頭尾相加為1、2、3等差。

30. B

解析：

思路一：124 是 1、2、4；3612是3、6、12；51020是5、10、20；71428是7、14、28；每列都成等差。

思路二：124，3612，51020，(71428)把每項拆成3個部分：[1,2,4]、[3,6,12]、[5,10,20]、[7,14,28]，每個[ ]中的新數列成等比。

思路三：首位數分別是1、3、5、（7），第二位數分別是：2、6、10、（14）；最後位數分別是：4、12、20、（28），故應該是71428，選B。

PART ONE
題庫練習
PART TWO
模擬試卷
PART THREE
考生急症室

(五)Interpretation of Tables & Graphs

Interpretation of Tables and Graphs (8 questions)

Directions :

This is a test on reading and interpretation of data presented in tables and graphs.

Examples:

Unemployment by Age Group

Age group	Year								
	1992	1993	1994	1995	1996	1997	1998	1999	2000
15 - 19	8.8 (6.6%)	7.5 (5.7%)	5.5 (4.4%)	7.4 (6.2%)	9.2 (8.1%)	6.8 (6.6%)	7.7 (8.1%)	7.2 (8.5%)	10.2 (12.7%)
20 - 29	20.2 (2.1%)	17.5 (1.9%)	14.5 (1.6%)	15.8 (1.8%)	19.3 (2.3%)	22.3 (2.7%)	22.7 (2.7%)	21.3 (2.5%)	33.5 (4.0%)
30 - 39	8.1 (1.1%)	6.1 (0.8%)	4.4 (0.6%)	5.9 (0.7%)	9.5 (1.1%)	12.0 (1.4%)	11.9 (1.3%)	13.1 (1.3%)	23.4 (2.3%)
40 - 49	4.1 (1.0%)	3.1 (0.7%)	2.4 (0.5%)	3.9 (0.8%)	6.7 (1.3%)	7.4 (1.3%)	7.3 (1.3%)	9.0 (1.4%)	18.5 (2.7%)
50 - 59	4.6 (1.4%)	2.7 (0.8%)	2.2 (0.7%)	2.9 (1.0%)	4.5 (1.5%)	4.9 (1.7%)	5.6 (1.8%)	5.4 (1.8%)	9.8 (3.2%)
60 or over	1.5 (0.9%)	0.9 (0.6%)	0.6 (0.4%)	0.8 (0.5%)	1.0 (0.6%)	1.3 (0.8%)	1.6 (1.2%)	1.3 (1.0%)	2.2 (1.7%)
Overall	47.3 (1.7%)	37.8 (1.4%)	29.6 (1.1%)	36.7 (1.3%)	50.2 (1.8%)	54.7 (1.9%)	56.8 (2.0%)	57.3 (1.9%)	97.6 (3.2%)

PART ONE
題庫練習

PART TWO
模擬試卷

PART THREE
考生急症室

Note:

(1) Figures in bold type are the numbers of unemployed persons in thousands in the labour force and figures in brackets are the corresponding unemployment rates.

(2) Unemployment rate refers to the proportion of unemployed persons in the labour force.

It is calculated as:

(number of unemployed persons of a particular age group / size of labour force of the same age group) x 100%

1. **During the period from 1992 to 2000, which age group consistently had the highest unemployment rate?**

   A. 15-19

   B. 20-29

   C. 30-39

   D. 40-49

   E. 50-59

Answer: A

Percentages in the brackets were the corresponding unemployment rates. By comparing the rates of different age groups in the same year, the age group of 15-19 was consistently the highest. The answer is A.

2. Estimate the total number of persons in the labour force in 1999.

    A. 1,090,000

    B. 3,020,000

    C. 3,050,000

    D. 3,130,000

    E. 5,720,000

Answer: B

The total number of unemployed persons in 1999 was 57,300 which constituted 1.9% of the overall labour force. The total number of persons in the labour force was 57,300÷1.9%=3,015,789. The answer is B.

PART ONE
題庫練習

PART TWO
模擬試卷

PART THREE
考生急症室

Question 1 to 2

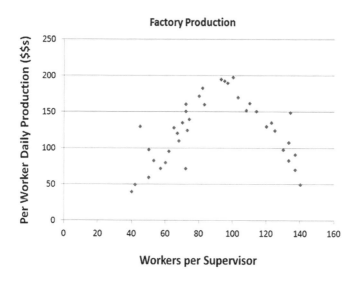

1. Assuming the trend in the graph stays the same over the range of workers per supervisor values, if the company were to employ 20 workers per supervisor they would likely be _____ productive than if the company were to employ 160 workers per supervisor.

   A. Less
   B. More
   C. Equally
   D. Not mentioned

2. If it wanted to increase productivity, a company currently employing 115 workersper supervisor should consider:

A. adding 15% more workers
B. halving the number of supervisors
C. doubling the number of supervisors
D. reducing the worker staff by 25%

Question 3 to 4

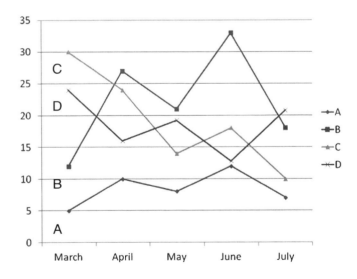

The above graph gives the values for 4 items measured by the police department. One represents the number of crimes reported, one repre-

PART ONE
題庫練習

PART TWO
模擬試卷

PART THREE
考生急症室

sents the number of arrests made, one represents the number of police officers on staff, and one represents the budget surplus for the department (in $1000s).

3. **If the impact of a change enacted by the police department takes a month to register, then we can identify a negative linear relationship between B and:**

   A. A
   B. -
   C. C
   D. D

4. **Given that each new police officer hired will make multiple arrests, and each new hire will immediately cut into the budget, _____ represents the department's budget surplus.**

   A. A
   B. B
   C. C
   D. D

Question 5 to 9

Study the following table and answer the questions based on it:

Year	Item of Expenditure				
	Salary	Fuel and Transport	Bonus	Interest on Loans	Taxes
1998	288	98	3.00	23.4	83
1999	342	112	2.52	32.5	108
2000	324	101	3.84	41.6	74
2001	336	133	3.68	36.4	88
2002	420	142	3.96	49.4	98

Expenditures of a Company (in Pesetas) per Annum Over the given Years.

5. **What is the average amount of interest per year which the company had to pay during this period?**

   A. 32.43

   B. 33.72

   C. 34.18

   D. 36.66

PART ONE
題庫練習
PART TWO
模擬試卷
PART THREE
考生急症室

6. The total amount of bonus paid by the company during the given period is approximately what percent of the total amount of salary paid during this period?

    A. 0.1%

    B. 0.5%

    C. 1.0%

    D. 1.25%

7. Total expenditure on all these items in 1998 was approximately what percent of the total expenditure in 2002?

    A. 62%

    B. 66%

    C. 69%

    D. 71%

8. The total expenditure of the company over these items during the year 2000 is?

    A. 544.44

    B. 501.11

    C. 446.46

    D. 478.87

9. The ratio between the total expenditure on Taxes for all the years and the total expenditure on Fuel and Transport for all the years respectively is approximately?

A. 4:7
B. 10:13
C. 15:18
D. 5:8

Question 10 to 15

Study the following table and answer the questions.Number of Candidates Appeared and Qualified in a Competitive Examination from Different States Over the Years.

State	Year									
	1997		1998		1999		2000		2001	
	App.	Qual.	App.	Qual.	App.	Qual.	App.	Qual.	App.	Qual.
M	5200	720	8500	980	7400	850	6800	775	9500	1125
N	7500	840	9200	1050	8450	920	9200	980	8800	1020
P	6400	780	8800	1020	7800	890	8750	1010	9750	1250
Q	8100	950	9500	1240	8700	980	9700	1200	8950	995
R	7800	870	7600	940	9800	1350	7600	945	7990	885

10. Total number of candidates qualified from all the states together in 1997 is approximately what percentage of the total number of candidates qualified from all the states together in 1998?

    A. 72%
    B. 77%
    C. 80%
    D. 83%

11. What is the average candidates who appeared from State Q during the given years?

    A. 8700
    B. 8760
    C. 8990
    D. 8920

12. In which of the given years the number of candidates appeared from State P has maximum percentage of qualified candidates?

    A. 1997
    B. 1998
    C. 1999
    D. 2001

13. What is the percentage of candidates qualified from State N for all the years together, over the candidates appeared from State N during all the years together?

   A. 12.36%

   B. 12.16%

   C. 11.47%

   D. 11.15%

14. The percentage of total number of qualified candidates to the total number of appeared candidates among all the five states in 1999 is?

   A. 11.49%

   B. 11.84%

   C. 12.21%

   D. 12.57%

15. Combining the states P and Q together in 1998, what is the percentage of the candidates qualified to that of the candidate appeared?

   A. 10.87%

   B. 11.49%

   C. 12.35%

   D. 12.54%

PART ONE
題庫練習

PART TWO
模擬試卷

PART THREE
考生急症室

Question 16 to 20

Sales of Books (in thousand numbers) from Six Branches - B1, B2, B3, B4, B5 and B6 of a publishing Company in 2000 and 2001.

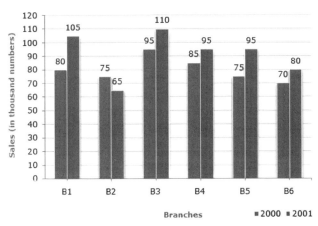

The bar graph given below shows the sales of books (in thousand number) from six branches of a publishing company during two consecutive years 2000(left bar) and 2001(right bar).

16. What is the ratio of the total sales of branch B2 for both years to the total sales of branch B4 for both years?

   A. 2:3
   B. 3:5
   C. 4:5
   D. 7:9

17. Total sales of branch B6 for both the years is what percent of the total sales of branches B3 for both the years?

A. 68.54%

B. 71.11%

C. 73.17%

D. 75.55%

18. What percent of the average sales of branches B1, B2 and B3 in 2001 is the average sales of branches B1, B3 and B6 in 2000?

A. 75%

B. 77.5%

C. 82.5%

D. 87.5%

19. What is the average sales of all the branches (in thousand numbers) for the year 2000?

A. 73

B. 80

C. 83

D. 88

20. Total sales of branches B1, B3 and B5 together for both the years (in thousand numbers) is?

A. 250

B. 310

C. 435

D. 560

Question 21 to 25

The bar graph given below shows the foreign exchange reserves of a country (in million US $) from 1991 - 1992 to 1998 - 1999.

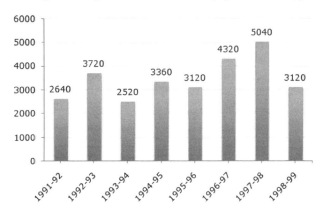

Foreign Exchange Reserves Of a Country. (in million US $)

The bar graph given shows the foreign exchange reserves of a country (in million US $) from 1991-1992 to 1998-1999

21. **The ratio of the number of years, in which the foreign exchange reserves are above the average reserves, to those in which the reserves are below the average reserves is?**

    A. 2:6

    B. 3:4

    C. 3:5

    D. 4:4

22. The foreign exchange reserves in 1997-98 was how many times that in 1994-95?

    A. 0.7
    B. 1.2
    C. 1.4
    D. 1.5

23. For which year, the percent increase of foreign exchange reserves over the previous year, is the highest?

    A. 1992-93
    B. 1993-94
    C. 1994-95
    D. 1996-97

24. The foreign exchange reserves in 1996-97 were approximately what percent of the average foreign exchange reserves over the period under review?

    A. 95%
    B. 110%
    C. 115%
    D. 125%

25. What was the percentage increase in the foreign exchange reserves in 1997-98 over 1993-94?

A. 100

B. 150

C. 200

D. 620

PART **ONE**
題庫練習

PART TWO
模擬試卷

PART THREE
考生急症室

Question 26 to 30

The following pie-chart shows the percentage distribution of the expenditure incurred in publishing a book. Study the pie-chart and the answer the questions based on it.

Various Expenditures (in percentage) Incurred in Publishing a Book

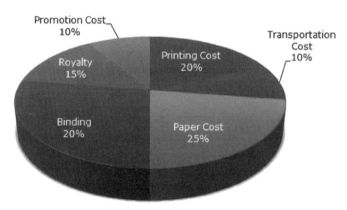

Various expenditures (in percentage) incurred in publishing a book

26. If for a certain quantity of books, the publisher has to pay Rs. 30,600 as printing cost, then what will be amount of royalty to be paid for these books?

A. Rs. 19,450
B. Rs. 21,200
C. Rs. 22,950
D. Rs. 26,150

27. What is the central angle of the sector corresponding to the expenditure incurred on Royalty?

A. 15€

B. 24€

C. 54€

D. 48€

28. The price of the book is marked 20% above the C.P. If the marked price of the book is Rs. 180, then what is the cost of the paper used in a single copy of the book?

A. Rs. 36

B. Rs. 37.50

C. Rs. 42

D. Rs. 44.25

29. If 5500 copies are published and the transportation cost on them amounts to Rs. 82500, then what should be the selling price of the book so that the publisher can earn a profit of 25%?

A. Rs. 187.50

B. Rs. 191.50

C. Rs. 175

D. Rs. 180

PART ONE
題庫練習

PART TWO
模擬試卷

PART THREE
考生急症室

30. Royalty on the book is less than the printing cost by:

    A. 5%

    B. 33 1/5%

    C. 20%

    D. 25%

Question 31 to 35

The following line graph gives the ratio of the amounts of imports by a company to the amount of exports from that company over the period from 1995 to 2001.

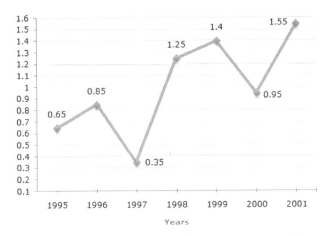

Ratio of value of imports to exports by a company over the years

31. If the imports in 1998 was Rs. 250 and the total exports in the years 1998 and 1999 together was Rs. 500 then the imports in 1999 was?

   A. 250
   B. 300
   C. 357
   D. 420

PART ONE
題庫練習
PART TWO
模擬試卷
PART THREE
考生急症室

32. The imports were minimum proportionate to the exports of the company in the year?

    A. 1995
    B. 1996
    C. 1997
    D. 2000

33. What was the percentage increase in imports from 1997 to 1998?

    A. 72
    B. 56
    C. 28
    D. Data inadequate

34. If the imports of the company in 1996 was Rs. 2 720 000 000 , the exports from the company in 1996 was?

    A. 3,700,000,000
    B. 3,200,000,000
    C. 2,800,000,000
    D. 2,750,000,000

35. In how many of the given years werethe exports more than the imports?

A. 1

B. 2

C. 3

D. 4

PART **ONE**
題庫練習

PART TWO
模擬試卷

PART THREE
考生急症室

答案及解析：

1. B

Explanation: Taking a close look at our curve, we can see that it's not exactly a bell-curve (with equal rates of decline on either side of the peak) - the rise on the left hand side of the curve is slower than the decline on the right-hand side of the curve. The peak takes place around 100 workers per supervisor, yet 40 workers per supervisor (60 away from the peak value) are just about equally productive as 140 workers per supervisor (40 away from the peak value). If we were to extend a line on the left side of the graph, we will find that at about 30 workers per supervisor, per worker daily production is zero. If we were to extend the trend on the right side of the graph, we will find that at about 155 workers per supervisor,per worker daily production is zero. This suggests that at 160 workers per supervisor, per worker daily production will be negative. Therefore, the company will likely be more productive at 30 workers per supervisor than at 160 workers per supervisor. Choose (B).

Note that extrapolating beyond the range of the data is generally not a good idea and could lead to some nonsensical conclusions. However, since you were asked to assume that the trend stays the same over the range of workers per supervisor values, we can infer that the company will be more productive at 30 workers per supervisor than at 160 workers per supervisor.

2. D

Explanation: At 115 workers per supervisor, we are on the sloping down side of the curve, suggesting that we need to have slightly fewer workers per supervisor. The question asks about which option would increase productivity (not necessarily maximize it, just increase it), and since only one option can be right, it must be that three of the options will not increase productivity and only one will.

First option - Adding more workers will mean moving to the right on the curve - more workers per supervisor - which clearly has lower productivity.

Second option - halving the number of supervisors would mean doubling the number of workers per supervisor, to 230, and as per the trend of the graph, this would presumably have lower productivity.

Third option - doubling the number of supervisors would mean that the ratio of workers to supervisor would be halved, to around 57.5 workers per supervisor, which has a lower productivity rate than 115 workers per supervisor.

Fourth option - with 25% fewer workers, the company would have around 87 workers per supervisor. Looking at the curve, although the difference is not huge, this is clearly a more productive ratio than 115 workers per supervisor. Select this final option.

3. C

Explanation: If there is a negative linear relationship, that means that when one variable goes up, the other goes down, and vice versa. The new wrinkle introduced by this question is that there can be a one-month lag in seeing this effect. B and D are opposites - when one goes up, the other goes down. If we were to introduce a one-month lag then it's possible they could correlate directly, not indirectly. With a one-month lag, B and A now seems to be have a negative relationship - when one goes up, the other goes down the following month, and vice versa. But is it a linear relationship - can it be expressed in the form A = mB +c, for some constant values m and c, given a month's lagtime? You don't have to do the math, just look at the graph - in the first month, B went up a lot, and in the second month, A went down a little. In the second month, B went down a little, yet in the third month A went up a lot. This is inconsistent - if there's a linear relationship, then the relative sizes of the changes need to be consistent with each other. Even if you look at the lag in the reverse direction, you'll notice this same inconsistency.

At this point, you could mark C by process of elimination, but let's just make sure it works. C shows a decrease in the first 2 months, while B constantly alternates between increases and decreases. So let C lag one month behind B. Between March and April, B has a big increase, and between April and May, C has a relatively large decrease. Between April and May, B has

PART ONE
題庫練習
PART TWO
模擬試卷
PART THREE
考生急症室

a smaller decrease, and between May and June, C has a smaller increase. Between May and June, B has an increase larger than the decrease but not as large as the first increase, and between June and July, C has a decrease larger than the previous increase but not as large as the first decrease. C is the correct option.

4. D

Explanation: If each new police officer makes multiple arrests, then the relationship between police officers on staff and arrests will be positive, and the total number of arrests will be greater than the total number of police officers. Looking at our graph, that could only mean that A is # of police officers and B is # of arrests. Every new police officer hired lowers the budget surplus, so there will be a negative relationship between A and the variable representing the surplus. Since C and A both decrease between April and May, this can only be D. Select the final option, D.

5. D

Explanation:

Average amount of interest paid by the Company during the given period

= (23.4 + 32.5 + 41.6 + 36.4 + 49.4)/5

= (183.3)/5

= 36.66

6. C

Explanation:

[(3.00 + 2.52 + 3.84 + 3.68 + 3.96)/(288 + 342 + 324 + 336 + 420)] x 100%

= (17/1710) x 100%

= approximately 1%

7. C

Explanation:

[(288 + 98+ 3.00 + 23.4 + 83)/(420 + 142 + 3.96 + 49.4 + 98)] x 100%

= (495.4/713.36) x 100%

= 69.45%

8. A

Explanation:

Total expenditure of the Company during 2000

= (324 + 101 + 3.84 + 41.6 + 74) pesetas

= 544.44 pesetas

9. B

Explanation:

[(83 + 108+ 74 + 88 + 98)/(98 + 112 + 101 + 133 + 142)]

= 451/586

= 1/1.3

= 10/13

10. C

Explanation:

[(720 + 840 + 780 + 950 + 870)/(980 + 1050 + 1020 + 1240 + 940)]x100%

= (4160/5230) x 100%

= 79.54%

= Approximately 80%

PART ONE
題庫練習

PART TWO
模擬試卷

PART THREE
考生急症室

11. C

Explanation:

(8100 + 9500 + 8700 + 9700 + 8950)/5

= 44950/5

= 8990

12. D

Explanation:

For 1997, (780/6400 x 100)% = 12.19%

For 1998, (1020/8800 x 100)% = 11.59%

For 1999, (890/7800 x 100)% = 11.41%

For 2000, (1010/8750 x 100)% = 11.54%

For 2001, (1250/9750 x 100)% = 12.82%

Maximum percentage is for the year 2001

13. D

Explanation:

[(840 + 1050 + 920 + 980 + 1020)/(7500 + 9200 + 8450 + 9200 + 8800)] x 100%

= (4810/43150) x 100%

= 11.15%

14. B

Explanation:

[(850 + 920 + 890 + 980 + 1350)/(7400 + 8450 + 7800 + 8700 + 9800)] x 100%

= (4990/42150) x 100%

= 11.84%

15. C

Explanation:

[(1020 + 1240)/(8800 + 9500)] x100%

= (2260/18300) x 100%

= 12.35%

16. D

Explanation:

Required ratio = (75 + 65)/(85 + 95)

= 140/180

= 7/9

17. C

Explanation:

Required ratio = [(70 + 80)/(95 + 110) x 100]%

= (150/205) x 100%

= 73.17%

18. D

Explanation:

Average sales (in thousand number) of branches B1, B3 and B6 in 2000

= 1/3 x (80 + 95 + 70) = (245/3)

Average sales (in thousand number) of branches B1, B2 and B3 in 2001

= 1/3 x (105 + 65 + 110) = (280/3)

Required percentage = [(245/3)/(280/3)] x 100%

= (245/280) x 100%

= 87.5%

PART ONE
題庫練習

PART TWO
模擬試卷

PART THREE
考生急症室

19. B

Explanation:

Average sales of all the six branches (in thousand numbers) for the year 2000

= 1/6 x (80 + 75 + 95 + 85 + 75 + 70)

= 80

20. D

Explanation:

Total sales of branches B1, b2 and B5 for both the years (in thousand numbers)

= (80 + 105) + (95 + 110) + (75 + 95)

= 560

21. C

Explanation:

Average foreign exchange reserves over the given period = 3480 million US $.

The country had reserves above 3480 million US $ during the years 1992-93, 1996-97 and 1997-98, i.e., for 3 years and below 3480 million US $ during the years 1991-92, 1993-94, 1994-95, 1995-56 and 1998-99, i.e., for 5 years.

Hence, required ratio = 3:5

22. D

Explanation:

Required ratio = 5040/3360

= 1.5

23. A

Explanation:

There is an increase in foreign exchange reserves during the years 1992-1993, 1994-1995, 1996-1997, 1997-1998 as compared to previous year (as shown by bar-graph).

The percentage increase in reserves during these years compared to previous year are:

For 1992-1993 = [(3720-2640)/2640] x 100% = 40.91%

For 1994-1995 = [(3360-2520)/2520] x 100% = 33.33%

For 1996-1997 = [(4320-3120)/3120] x 100% = 38.46%

For 1997-1998 = [(5040-4320)/4320] x 100% = 16.67%

Clearly, the percentage increase over previous year is highest for 1992-1993.

24. D

Explanation:

Average foreign exchange reserves over the given period

= [1/8 x (2640 + 3720 + 2520 + 3360 + 3120 + 4320 + 5040 + 3120)] million US $

= 3480 million US $

Foreign exchange reserves in 1996 - 1997 = 4320 million US $

Therefore, required percentage

= (4320/3480 x 100)%

= 124.14%

= approximately 125%

25. A

Explanation:

Foreign exchange reserves in 1997-1998 = 5040 million US $

Foreign exchange reserves in 1993-1994 = 2520 million US $

Therefore Increase = (5040-2520) = 2520 US $

Therefore Percentage Increase = (2520/2520 x 100)% = 100%

26. C

Explanation:

Let the amount of Royalty to be paid for these books be Rs. r.

Then, 20:15 = 30600:r

r = Rs. (30600x15)/20

= Rs. 22,950

27. C

Explanation:

Central angle corresponding to Royalty = (15% of 360)€

= (15/100)x 360€

= 54€

28. B

Explanation:

Clearly, marked price of the book = 120% of C.P.

Also, cost of paper = 25% of C.P.

Let the cost of paper for a single book be Rs. n.

Then, 120:25 = 180:n

=> n = Rs. (25x180)/120

= Rs. 37.50

29. A

Explanation:

For the publisher to earn a profit of 25%, S.P. = 125% of C.P.

Also Transportation Cost = 10% of C.P.

Let the S.P. of 5500 books be Rs. x

Then, 10:125 = 82500:x

=> x = Rs.(125x82500)/10

= Rs. 1031250

Therefore S.P. of one book = Rs. 1031250/5500

= Rs. 187.50

30. D

Explanation:

Printing cost of book = 20% of C.P.

Royalty on book = 15% of C.P.

Difference = (20% of C.P.) - (15% of C.P.) = 5% of C.P.

Therefore, percentage difference

= (Difference / Printing Cost) x 100%

= (5% of C.P. / Printing Cost) x 100%

= 25%

31. D

Explanation:

The ratio of imports to exports for the years 1998 and 1999 are 1.25 and 1.40 respectively.

Let the exports in the year 1998 = x

Then, the exports in the year 1999 = (500 - x)

Therefore 1.25 = 250/x

PART ONE
題庫練習
PART TWO
模擬試卷
PART THREE
考生急症室

=> x = 250/1.25

= 200 (Using ratio for 1998)

Thus, the exports in the year 1999 = (500 - 200) = 300

Let the imports in the year 1999 = y

Then, 1.40 = y/300 => y = (300 x 1.40) = 420

Therefore, imports in the year 1999 = 420

32. C

Explanation:

The imports are minimum proportionate to the exports implies that the ratio of the value of imports to exports has the minimum value.

Now, this ratio has a minimum value 0.35 in 1997, i.e., the imports are minimum proportionate to the exports in 1997.

33. D

Explanation:

The graph gives only the ratio of imports to exports for different years. To find the percentage increase in imports from 1997 to 1998, we require more details such as the value of imports or exports during these years.

Hence, the data is inadequate to answer this question.

34. B

Explanation:

Ratio of imports to exports in the year 1996 = 0.85

Let the exports in 1996 = x

Then, 2720000000/x = 0.85

=> x = 2,720,000,000/0.85 = 320

Therefore, exports in 1996 = 3,200,000,000

35. D

Explanation:

The exports are more than the imports imply that the ratio of value of imports to exports is less than 1.

Now, this ratio is less than 1 in years 1995, 1996, 1997 and 2000.

Thus, there are four such years.

模擬試卷一

能力傾向測試

模擬測驗（一）

限時四十五分鐘

PART ONE
題庫練習

PART **TWO**
模擬試卷

PART THREE
考生急症室

（一）演繹推理（8題）

請根據以下短文的內容，選出一個或一組推論。請假定短文的內容都是正確的。

1. 產品定位是指企業為了滿足目標市場，確定產品（或服務）的功能、質量、價格、包裝、銷售渠道、服務方式等。

   根據以上定義，下列不屬於產品定位的是：

   A. 我們的產品將為在校學生服務

   B. 考慮到消費人群的收入狀況，我們的產品將定價在1000元上下

   C. 我們的服務主要將通過上門維修來實現

   D. 我們的服務會讓你有美的享受

2. 美國是全球有機豬肉產量最大的國家，但其價格卻只比普通豬肉高兩三成。在中國，有機豬肉竟比普通豬肉的價格高好幾倍。因此，中國的有機豬肉行業是暴利行業。

   上述結論如果正確，需要以下哪項假設？

   A. 中國人對食品安全的普遍擔憂導致有機豬肉供不應求

   B. 中國普通豬肉的價格完全是市場化的，其利潤是正常的

   C. 中國有機豬肉的品種比美國的多

   D. 中國的有機豬肉不比美國的有機豬肉成本高，普通豬肉價格也相當

3. 國家賠償是指國家機關及其工作人員違法行使行政、偵察、檢查、審判、監獄管理等職權，侵犯公民、法人和其他組織的合法權益並造成損害的，由法律規定的賠償義務機關對受害人予以賠償的法律制度。

根據以上定義，下列屬於國家賠償的是：

A. 某戰士因公殉職，國家追認其為烈士，並發給其家人每年1萬元的撫恤金

B. 公務員李先生因丟失從鄰居處借來的一條金項鏈，賠了人家3000多元

C. 工商機關因錯誤地吊銷某個體戶的營業執照，使其蒙受經濟損失，因此給予單次過2萬元的賠償

D. 國家財政每年從稅收中拿出一部分用於扶貧建設

4. 刻板印象是指人們通過直接或間接經驗形成起來的，對社會中某一群體或事件較為固定的看法。

根據以上定義，下列不屬於刻板印象的是：

A. 商人是唯利是圖的

B. 娛樂圈就是烏七八糟的

C. 北方人是豪爽的

D. 靜止是相對的

PART ONE
題庫練習

PART TWO
模擬試卷

PART THREE
考生急症室

5. 法院調解就是指在人民法院審判人員主持下，雙方當事人自願就民事權益爭議平等協商，達成協議，從而解決糾紛所進行的活動。

根據以上定義，下列屬於法院調解的是：

A. 某地區林業機關與農業機關因權限交叉產生衝突，告之法庭，後在上級機關的調解下達成協議並撤訴

B. 兩家因一件小事發生糾紛，打起了漫長的官司，花費巨大，後都感到不值得，於是私下和解後紛紛撤訴

C. 兩夫妻一方提出要離婚，另一方則拿不定主意，後法院審判員在對他們講明是非的基礎上使他們又和好如初

D. 兩姐妹合開一家理髮館，後因分紅問題發生糾紛，但在她們即將鬧上法庭的時候，終於達成諒解

6. 回避條件作用是指當厭惡、刺激或不愉快的情景出現時，個人作出某種反應，從而逃避了厭惡、刺激或不愉快情景，則該反應在以後的類似情景中發生的概率增加。

根據以上定義，下列不屬於回避條件作用的一項是：

A. 因為貪睡不願起來吃早餐

B. 碰到熱燙的東西，趕緊縮回手來

C. 感覺屋內人聲嘈雜時暫時離開

D. 看見地上的垃圾後繞開走

7. 美國《科學》雜誌公布了本年度科學研究十大攻破之一是：通過對人和大鼠的研究，科學家提出，記憶力和想像力均植根於人類大腦的海馬區，該區是大腦重要的記憶中心。

由此推斷：

A. 大腦記憶也許能重新梳理過去的經歷

B. 大腦記憶也許能產生關於未來的想像

C. 科學家發現人類大腦的重要記憶中心

D. 科學家發現挑戰人類智力的記憶程序

8. 摩擦性失業是勞動者正常流動或生產中不可避免的摩擦造成的短期、局部性失業。

根據以上定義，以下屬於摩擦性失業的是：

A. 由於冬季是旅遊淡季，導遊張先生被公司以節省開支為名炒了魷魚

B. 李女士在生完孩子後還沒有找到新的工作

C. 劉先生放棄了在現在的工作，跑到別處去發展，但還沒有找到合適的工作

D. 王先生剛大學畢業，還沒有找到自己滿意的工作

（二）Verbal Reasoning

（6 questions）

In this test, each passage is followed by three statements (the questions). You have to assume what is stated in the passage is true and decide whether the statements are either:

True (Box A): The statement is already made or implied in the passage, or follows logically from the passage.

False (Box B): The statement contradicts what is said, implied by, or follows logically from the passage.

Can't tell (Box C): There is insufficient information in the passage to establish whether the statement is true or false.

Passage 1 (Question 9 to 11)

Crude oil (also known as petroleum) is a type of fossil fuel found beneath the earth's surface. It is formed by the gradual build-up of fossilised organic materials such as algae and plankton. As more layers build up, the bottom most layers are heated and subject to pressure, with the combination of heat and pressure leading to the matter transforming into the waxy substance kerogen. Following even more prolonged exposure to heat and pressure, the kerogen eventually becomes transformed into liquid and gases via the catagenesis process. The formation of crude oil occurs from this pyrolysis (heating) process. The range of heat at which kerogen becomes crude oil is called the oil window. Below this range the crude oil remains kerogen and above this point the crude oil becomes a natural gas.

9. If the temperature is too low, crude oil remains in a solid state, whereas if it is too hot, it becomes a gas.

10. Crude oil is non-renewable.

11. Kerogen becomes crude oil after further heating and pressurisation in the catagenesis process.

PART ONE
題庫練習

PART **TWO**
模擬試卷

PART THREE
考生急症室

Passage 2 (Question 12 to 14)

Wine is an alcoholic beverage made from fermented grapes. Wine is often associated most closely with France and with some justification. France is easily the biggest producer of wine, biggest consumer (on a per capita basis) and usually the biggest exporter of it. The process of creating wine is called vinification or winemaking. Grapes are crushed and then fermented using yeast. Yeast has the effect of consuming the sugars in grapes and then producing alcohol. Carbon dioxide is also produced as part of this process, but it is normally not captured. There are additional processes depending on the type of wine that is to be produced. For example, red wine undergoes a secondary fermentation which includes the conversion of malic acid into lactic acid, which aims to soften the taste of the wine. Red wine may also be transferred to oak barrels for maturity to induce an 'oakiness' on the produce.

12. Carbon dioxide, which is a by-product of the winemaking process, is sometimes captured and then used to carbonate other beverages.

13. White wine may be transferred to oak barrels to create an 'oakiness' flavor dimension.

14. Vinification is the same thing as winemaking.

（三）Data Sufficiency Test

（8 questions）

In this test, you are required to choose a combination of clues to solve a problem.

15. Is 13N a positive number?

(1) -21N is a negative number

(2) N(2)<1

A. Statement (1) ALONE is sufficient, but statement (2) alone is not sufficient.

B. Statement (2) ALONE is sufficient, but statement (1) alone is not sufficient.

C. BOTH statements TOGETHER are sufficient, but NEITHER statement ALONE is sufficient.

D. EACH statement ALONE is sufficient.

E. Statement (1) and (2) TOGETHER are NOT sufficient to answer the question asked, and additional data are needed.

16. What is the average (arithmetic mean) of w, x, y, z, and 10?

(1) the average (arithmetic mean) of w and y is 7.5; the average (arithmetic mean) of x and z is 2.5

(2) -[-z-y-x-w]=20

PART ONE
題庫練習

PART **TWO**
模擬試卷

PART THREE
考生急症室

A. Statement (1) ALONE is sufficient, but statement (2) alone is not sufficient.

B. Statement (2) ALONE is sufficient, but statement (1) alone is not sufficient.

C. BOTH statements TOGETHER are sufficient, but NEITHER statement ALONE is sufficient.

D. EACH statement ALONE is sufficient.

E. Statement (1) and (2) TOGETHER are NOT sufficient to answer the question asked, and additional data are needed.

17. **If X is a positive integer, is X a prime number?**

   (1) X is an even number

   (2) 1<X<4

A. Statement (1) ALONE is sufficient, but statement (2) alone is not sufficient.

B. Statement (2) ALONE is sufficient, but statement (1) alone is not sufficient.

C. BOTH statements TOGETHER are sufficient, but NEITHER statement ALONE is sufficient.

D. EACH statement ALONE is sufficient.

E. Statement (1) and (2) TOGETHER are NOT sufficient to answer the question asked, and additional data are needed.

18. **What is the remainder of a positive integer N when it is divided by 2?**

    (1) N contains odd numbers as factors

    (2) N is a multiple of 15

    A. Statement (1) ALONE is sufficient, but statement (2) alone is not sufficient.
    B. Statement (2) ALONE is sufficient, but statement (1) alone is not sufficient.
    C. BOTH statements TOGETHER are sufficient, but NEITHER statement ALONE is sufficient.
    D. EACH statement ALONE is sufficient.
    E. Statement (1) and (2) TOGETHER are NOT sufficient to answer the question asked, and additional data are needed.

19. **What is the value of $(n+1)^2$?**

    (1) $N^2-6n=-9$

    (2) $(n-1)^2=n^2-5$

    A. Statement (1) ALONE is sufficient, but statement (2) alone is not sufficient.
    B. Statement (2) ALONE is sufficient, but statement (1) alone is not sufficient.
    C. BOTH statements TOGETHER are sufficient, but NEITHER statement ALONE is sufficient.

PART ONE
題庫練習

PART TWO
模擬試卷

PART THREE
考生急症室

D. EACH statement ALONE is sufficient.

E. Statement (1) and (2) TOGETHER are NOT sufficient to answer the question asked, and additional data are needed.

20. $15a+6b=30$, what is the value of a-b?

(1) $b=5-2.5a$

(2) $9b=9a-81$

A. Statement (1) ALONE is sufficient, but statement (2) alone is not sufficient.

B. Statement (2) ALONE is sufficient, but statement (1) alone is not sufficient.

C. BOTH statements TOGETHER are sufficient, but NEITHER statement ALONE is sufficient.

D. EACH statement ALONE is sufficient.

E. Statement (1) and (2) TOGETHER are NOT sufficient to answer the question asked, and additional data are needed.

21. X, Y, and Z are three points in space; is Y the midpoint of XZ?

(1) ZY and YX have the same length

(2) XZ is the diameter of a circle with center Y

A. Statement (1) ALONE is sufficient, but statement (2) alone is not sufficient.

B. Statement (2) ALONE is sufficient, but statement (1) alone is not sufficient.

C. BOTH statements TOGETHER are sufficient, but NEITHER statement ALONE is sufficient.

D. EACH statement ALONE is sufficient.

E. Statement (1) and (2) TOGETHER are NOT sufficient to answer the question asked, and additional data are needed.

22. How many members of the staff of Advanced Computer Technology Consulting are women from outside the United States?

(1) one-fourth of the staff at Advanced Computer Technology Consulting are men

(2) 20% of the staff, or 20 individuals, are men from the U.S.; there are twice as many women from the U.S. as men from the U.S.

PART ONE
題庫練習

PART **TWO**
模擬試卷

PART THREE
考生急症室

A. Statement (1) ALONE is sufficient, but statement (2) alone is not sufficient.
B. Statement (2) ALONE is sufficient, but statement (1) alone is not sufficient.
C. BOTH statements TOGETHER are sufficient, but NEITHER statement ALONE is sufficient.
D. EACH statement ALONE is sufficient.
E. Statement (1) and (2) TOGETHER are NOT sufficient to answer the question asked, and additional data are needed.

（四）Numerical Reasoning

（5 questions）

Each question is a sequence of numbers with one or two numbers missing. You have to figure out the logical order of the sequence to find out the missing number(s).

(23) 124，3612，51020，( )

- A. 77084
- B. 71428
- C. 81632
- D. 91386

(24) 1，1，2，3，5，8，( )

- A. 11
- B. 12
- C. 13
- D. 14

(25) 1，1，3，5，11，( )

- A. 8
- B. 13
- C. 21
- D. 32

PART ONE
題庫練習
PART **TWO**
模擬試卷
PART THREE
考生急症室

(26) 0，1，1，2，3，( )，22

A. 5
B. 7
C. 9
D. 11

(27) 3，4，9，28，113，( )

A. 566
B. 678
C. 789
D. 961

（五．）Interpretation of Tables & Graphs (8 questions)

This is a test on reading and interpretation of data presented in tables and graphs.

Graph 1 (Question 28 to 31)

The following table gives the percentage of marks obtained by seven students in six different subjects in an examination.

Student	Subject (Max. Marks)					
	Maths	Chemistry	Physics	Geography	History	Computer Science
	(150)	(130)	(120)	(100)	(60)	(40)
Ayush	90	50	90	60	70	80
Aman	100	80	80	40	80	70
Sajal	90	60	70	70	90	70
Rohit	80	65	80	80	60	60
Muskan	80	65	85	95	50	90
Tanvi	70	75	65	85	40	60
Tarun	65	35	50	77	80	80

The numbers in the brackets give the maximum marks in each subject

PART ONE
題庫練習

PART TWO
模擬試卷

PART THREE
考生急症室

28. What are the average marks obtained by all the seven students in Physics? (rounded off to two digit after decimal)

A. 77.26%
B. 89.14
C. 91.37
D. 96.11

29. The number of students who obtained 60% and above marks in all subjects is?

A. 1
B. 2
C. 3
D. none

30. What was the aggregate of marks obtained by Sajal in all the six subjects?

A. 409
B. 419
C. 429
D. 449

31. In which subject is the overall percentage the best?

A. Maths
B. Chemistry
C. Physics
D. History

Graph 2 (Question 32 to 35)

The bar graph given below shows the data of the production of paper (in lakh tonnes = One Lakh is equal to One Hundred Thousand (100,000)) by three different companies X, Y and Z over the years.

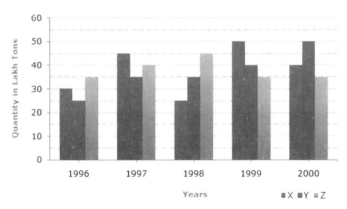

Production of paper (in lakh tonnes) by Three Companies X, Y and Z over the Years.

32. For which of the following years, the percentage rise/fall in production from the previous year is the maximum for Company Y?

A. 1997

B. 1998

C. 1999

D. 2000

PART ONE
題庫練習

PART TWO
模擬試卷

PART THREE
考生急症室

33. What is the ratio of the average production of Company X in the period 1998-2000 to the average production of Company Y in the same period?

    A. 1:1
    B. 15:17
    C. 23:25
    D. 27:29

34. **The average production for five years was maximum for which company?**

    A. X
    B. Y
    C. Z
    D. X and Z both

35. **In which year was the percentage of production of Company Z to the production of Company Y the maximum?**

    A. 1996
    B. 1997
    C. 1998
    D. 1999

END OF PAPER

ANSWER - PAPER 1

（一）演繹推理（8題）

1. D

解析：由定義能夠看出產品定位有六方面的內容：產品（或服務）的功能、質量、價格、包裝、銷售渠道、服務方式。A項屬於銷售渠道；B項屬於產品的價格；C項屬於服務方式；而D項是不包含在上面任何一項中的，因此不屬於產品定位。

2. D

解析：題幹的結論是中國的有機豬肉行業是暴利行業。D選項通過說明中國的有機豬肉的成本不比美國的有機豬肉高，普通豬肉價格也大致相同，肯定了中國有機豬肉的價格高會獲利多，是題幹需要的假設條件。故D項當選。

3. C

解析：A、B、D三項中均並不存在工作人員違法行使職權而造成公司、法人和其他組織的合法權益的損害的事實。C項中工商機關行使職權造成某個體戶合法權益遭受損害，故答案為C。

4. D

解析：D項不屬於對社會中某一群體或事件較為固定的看法，而是對客觀事物的科學看法。故答案選D。

5. C

解析：A項在上級機關的調解下達成協議，B項私下和解，D項鬧上法庭前達成諒解，均沒有在人民法院審判人員主持下進行調解。

6. A

解析：A項中並不是因為厭惡吃早餐，從而逃避吃早餐，而是由於貪睡不願

PART ONE
題庫練習

PART **TWO**
模擬試卷

PART THREE
考生急症室

起來吃早餐，故不屬於回避條件作用。

7. C

解析：

第一步：抓住題幹的對象並判斷整體關係

題幹闡述了科學發現了海馬區是大腦重要的記憶中心和想像中心。

第二步：逐一判斷選項的作用

A突出梳理過去，B突出想像未來，這在題幹中都未涉及到，所以A、B錯誤；根據題幹整體關係可知「海馬區是大腦重要的記憶中心和想像中心」，所以C正確；D強調發現記憶程序，而題幹強調發現了海馬區的記憶功能，所以D錯誤。故正確答案為C。

8. C

解析：A、B、D三項都是因個人因素失業，並不是勞動者正常流動或生產中不可避免的摩擦造成的短期、局部性失業。

（二）Verbal Reasoning (6 questions)

9. A (True)

Explanation: Kerogen is described as waxy substance (i.E. a solid). The final sentence notes that insufficient heat leaves it as kerogen (a solid), while more heat makes it natural gas (a gas).

10. C (Can't tell)

Explanation: There is no mention of this, although in reality it is true.

11. A (True)

Explanation: The 4th last sentence says this basically - '...following even more prolonged exposure to heat and pressure, the kerogen eventually becomes transformed into liquid...'.

12. C (Can't tell)

Explanation: Although carbon dioxide is mentioned as a bi-product of the winemaking process, there is no mention of capturing it for use in other products. Equally, there's nothing which explicitly excludes that possibility.

13. C (Can't tell)

Explanation: This is a tricky question which should teach you to be very careful when reading the passage and question. This is a classical verbal reasoning question which tempts you into selecting True, when in fact the answer is Can't Tell. The passage only mentions that the oakiness can be infused into red wine; it doesn't mention white wine. Be careful!

14. A (True)

Explanation: See sentence 4, which says exactly this.

(三) Data Sufficiency Test (8 questions)

15. A

Explanation: 1. Simplify the question:

Since multiplying a number by 13 does not change its sign, the question can be simplified to: "is N a positive number?"

2. Evaluate Statement (1) alone:

(a) Write out algebraically:

-21N=negative

21N=positive {divided by -1}

N=positive

(b) Since N is a positive number, 13N will always be a positive number.

(c) Statement (1) alone is SUFFICIENT.

3. Evaluate Statement (2) alone:

(a) Any time you are dealing with a number raised to an even exponent, you must remember that the even exponent hides the sign of the base (e.g., $x^2=16$; x=4 AND -4).

PART ONE
題庫練習

PART **TWO**
模擬試卷

PART **THREE**
考生急症室

(b) Solve the inequality:

$N^2 < 1$

$-1 < N < 1$ {take the square root, remembering that there is a positive and negative root}

(c) Since N can be both positive (e.g., .5) or negative (e.g., -.5), Statement (2) is not sufficient.

(d) Statement (2) alone is NOT SUFFICIENT.

4. Since Statement (1) alone is SUFFICIENT but Statement (2) alone is NOT SUFFICIENT, answer A is correct.

16. D

Explanation: 1. Write out the formula for the mean and arrange it in several different ways so that you can spot algebraic substitutions:

Mean=(w+x+y+z+10)/5

5*Mean=w+x+y+z+10

2. Evaluate Statement (1) alone:

(a) Translate each piece of information into algebra:

"the average (arithmetic mean) of w and y is 7.5"

(w+y)/2=7.5

w+y=15

"the average (arithmetic mean) of x and z is 2.5"

(x+z)/2=2.5

x+z=5

(b) Combine the two equations by adding them together:

(x+z)+(w+y)=(5)+(15)

x+z+w+y=15+5

w+x+y+z=20

(c) Substitute into the equation from the top:

Equation from top: 5*Mean=w+x+y+z+10

5*Mean=20+10=30

Mean=6

(d) Statement (1) alone is SUFFICIENT

3. Evaluate Statement (2) alone:

(a) Simplify the algebra:

-[-z-y-x-w]=20

z+y+x+w=20

(b) This can be substituted into the mean formula:

z+y+x+w=20

w+x+y+z=20{rearrange left side to make substitution easier to see}

Equation from top: 5*Mean=w+x+y+z+10

5*Mean=(w+x+y+z)+10

5*Mean=20+10=30{substitute information from Statement (2)}

Mean=6

(c) Statement (2) alone is SUFFICIENT.

4. Since Statement (1) alone is SUFFICIENT and Statement (2) alone is SUF-FICIENT, answer D is correct.

17. B

Explanation: 1. Evaluate Statement (1) alone

(a) Make a list of even numbers and evaluate whether they are prime:

2 - Prime

4 - Not Prime

6 - Not Prime

8 - Not Prime

10 - Not Prime

(b) Every single even number except 2 is not a prime number. However, since Statement (1) enables X to be prime (e.g., 2) and not prime (e.g., 4, 6, 8, 10, ...), Statement (1) is NOT SUFFICIENT.

(c) Statement (1) alone is NOT SUFFICIENT.

2. Evaluate Statement (2) alone.

(a) List the possible values of X, remembering that X is a positive integer such that $1<X<4$:

PART ONE
題庫練習

PART TWO
模擬試卷

PART THREE
考生急症室

X=2: Prime Number

X=3: Prime Number

Since all possible values of X given the parameters in Statement (2) are prime, Statement (2) is SUFFICIENT.

(b) Statement (2) alone is SUFFICIENT.

3. Since Statement (1) alone is NOT SUFFICIENT and Statement (2) alone is SUFFICIENT, answer B is correct.

18. E

Explanation: 1. Any positive integer that is divided by 2 will have a remainder of 1 if it is odd. However, it will not have a remainder if it is even.

N/2→Remainder=0 if N is even

N/2→Remainder=1 if N is odd

2. Evaluate Statement (1) alone:

(a) If a number contains only odd factors, it will be odd (and will have a remainder of 1 when divided by 2). If a number contains at least one even factor, it will be even (and divisible by 2).

15=3x5 (only odd factors; not divisible by 2; remainder of 1)

21=3x7 (only odd factors; not divisible by 2; remainder of 1)

63=3x3x7 (only odd factors; not divisible by 2; remainder of 1)

30=3x5x2 (contains an even factor; divisible by 2)

42=3x7x2 (contains an even factor; divisible by 2)

50=5x5x2 (contains an even factor; divisible by 2)

(b) Simply because "N contains odd numbers as factors" does not mean that all of N's factors are odd. Consequently, it is entirely possible that N contains an even factor, in which case N is even and N is divisible by 2. Possible values for N:

18=2x3x3 (contains odd factors, but is divisible by 2; remainder=0)

30=2x5x3 (contains odd factors, but is divisible by 2; remainder=0)

But:

27=3x3x3 (contains odd factors, but is not divisible by 2; remainder=1)

15=3x5 (contains odd factors, but is not divisible by 2; remainder=1)

(c) Since some values of N that meet the conditions of Statement (1) are divisible by 2 while other values that also meet the conditions of Statement (1) are not divisible by 2, Statement (1) does not provide sufficient information to definitively determine whether N is divisible by 2.

(d) Statement (1) alone is NOT SUFFICIENT.

3. Evaluate Statement (2) alone:

(a) Since "N is a multiple of 15", possible values for N include: 15, 30, 45, 60, 75, 90

(b) Possible values for N give different remainders when divided by 2:

15/2→Remainder=1

30/2→Remainder=0

45/2 →Remainder=1

60/2→Remainder=0

75/2→Remainder=1

90/2→Remainder=0

(c) Since different legitimate values of N give different remainders when divided by 2, Statement (2) is not sufficient for determining the remainder when N is divided by 2.

(d) Statement (2) alone is NOT SUFFICIENT.

4. Evaluate Statements (1) and (2):

(a) Since "N is a multiple of 15" and "N contains odd numbers as factors", possible values for N include: 15, 30, 45, 60, 75, 90

(b) Adding Statement (1) to Statement (2) does not provide any additional information since any number that is a multiple of 15 must also have odd numbers as factors.

(c) Possible values for N give different remainders when divided by 2:

15/2→Remainder=1

30/2→Remainder=0

45/2→Remainder=1

60/2→Remainder=0

75/2→Remainder=1

90/2→Remainder=0

PART ONE
題庫練習

PART **TWO**
模擬試卷

PART THREE
考生急症室

(d)Since different legitimate values of N give different remainders when divided by 2, Statements (1) and (2) are not sufficient for determining the remainder when N is divided by 2.

(e)Statements (1) and (2), even when taken together, are NOT SUFFICIENT.

5. Since Statement (1) alone is NOT SUFFICIENT and Statement (2) alone is NOT SUFFICIENT, answer E is correct.

19. D

Explanation: 1. Evaluate Statement (1) alone:

(a) Try to solve for n:

$n^2-6n=-9$

$n^2-6n+9=0$

$(n-3)^2=0$

$n-3=0$

$n=3$

With one value for n, we can find a single value for $(n + 1)^2$

(b) Statement (1) alone is SUFFICIENT.

2. Evaluate Statement (2) alone:

(a) Expand the terms and simplify them:

$n^2-2n+1=n^2-5$

$-2n+1=-5$

$-2n+6=0$

$6=2n$

$n=3$

With one value for n, we can find a single value for $(n+1)^2$

(b) Statement (2) alone is SUFFICIENT.

3. Since Statement (1) alone is SUFFICIENT and Statement (2) alone is SUFFICIENT, answer D is correct.

20. B

Explanation: 1. Be aware that simply because you have two equations with two unknowns does not mean that a solution exists. You must have two unique equations with two unknowns in order for a solution to exist.

2. Evaluate Statement (1) alone:

(a) There are two possible ways to solve this problem:

(i) Method (1): Substitute b from Statement (1) into the original equation.

$15a+6(5-2.5a)=30$

$15a+30-15a=30$

$30=30$

$0=0$

Based upon this answer, the equation in Statement (1) is the equation in the original question solved for B. Consequently, we only have one equation and two unknowns. There is not enough information to determine a-b.

(ii) Method (2): Rearrange the equation in Statement (1) and subtract this equation from the original equation.

$b=5-2.5a$

$b+2.5a=5$

$2.5a+b=5$

Multiply by 6 so b's cancel: $15a+6b=30$

This method also shows that the equation in Statement (1) is nothing more than the original equation rearranged. Consequently, we only have one equation and two unknowns. There is not enough information to determine a-b.

(b) Statement (1) is NOT SUFFICIENT.

3. Evaluate Statement (2) alone:

(a) Try to line up the two equations so that you can subtract them:

$9b=9a-81$

$81+9b=9a$

$81=9a-9b$

Statement (2) Equation: $9a-9b=81$

Original Question Equation: $15a+6b=30$

PART ONE
題庫練習

PART **TWO**
模擬試卷

PART THREE
考生急症室

At this point, you can stop since you know that you have two unique equations and two unknowns. Consequently, there will be a solution for a and for b, which means there will be one unique value for a- b. Statement (2) is SUFFICIENT.

(b) If you want to solve to see this (Note: Do not solve this in a test as it takes too much time and is not necessary):

Multiply (2) by 4:36a-36b=324

Multiply Original by 6: 90a+36b=180

6xOriginal+2xStatement(2):(90a+36a)+(36b+-36b)=180+324

126a=204

a=4

Solve for b:

9b=9(4)-81=-45

b=-5

a-b=4-(-5)=4+5=9

4. Since Statement (1) alone is NOT SUFFICIENT and Statement (2) alone is SUFFICIENT, answer B is correct.

21. B

Explanation: 1. Evaluate Statement (1) alone:

(a) It is possible that XZ is a straight line with Y as the midpoint, making ZY=YX.

(b) However, just because ZY=YX does not mean Y must always be the midpoint; XYZ could be an equilateral triangle.

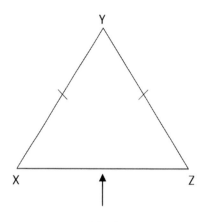

Midpoint

(c) Statement (1) alone is NOT SUFFICIENT.

2. Evaluate Statement (2) alone.

(a) By definition, the center of a circle is the midpoint of a diameter. Consequently, XZ runs through point Y and XY=YZ since both are radii and all radii must be the same length.

(b) Statement (2) alone is SUFFICIENT.

3. Since Statement (1) alone is NOT SUFFICIENT but Statement (2) alone is SUFFICIENT, answer B is correct.

22. C

Explanation: 1. Note that this question asks for a specific number, not a ratio. Consequently, keep in mind that knowing y percent of the total staff is composed of women from outside the United States is not sufficient.

PART ONE
題庫練習
PART **TWO**
模擬試卷
PART THREE
考生急症室

2. Evaluate Statement (1) alone:

(a) If 25% of the staff are men, 75% must be women.

(b)

	Men	Women	Total
From U.S.			
From Outside U.S.			
	.25(x)	.75(x)	x

(c) There is not enough information to determine the number of women from outside the United States. Statement (1) alone is NOT SUFFICIENT.

3. Evaluate Statement (2) alone:

(a) Since 20 men from the U.S. represent 20% of the staff, the total staff is 100. We also know that there are 20 men from the U.S. and 2(20)=40 women from the U.S. for a total of 20+40=60 employees from the U.S. Consequently, 100-60=40 employees must be from outside the U.S.

(b)

	Men	Women	Total
From U.S.	20	40	60
From Outside U.S.			40
			x=100

(c) Since we cannot determine the breakdown of the 40 employees from outside the U.S., it is impossible to determine the number of women from outside the U.S.; Statement (2) alone is NOT SUFFICIENT.

4. Evaluate Statements (1) and (2) together:

(a) Fill in as much information as possible from Statements (1) and (2). We now know that there are a total .25(x)=.25(100)=25 men and .75(x)=.75(100)=75 women.

(b)

	Men	Women	Total
From U.S.	20	40	60
From Outside U.S.	5	35	40
	25	75	x=100

(c) 35 members of the staff of Advanced Computer Technology Consulting are women from outside the United States.

5. Since Statement (1) alone is NOT SUFFICIENT and Statement (2) alone is NOT SUFFICIENT, but Statements (1) and (2), when taken together, are SUFFICIENT, answer C is correct.

（四）Numerical Reasoning (5 questions)

23. B

解析：

解法1：將數字進行機械劃分：1|2|4、3|6|12、5|10|20、(| |)。

每一項劃分成三部分，各自構成新數列。

第一部分：1、3、5、(7)，成等差數列；

第二部分：2、6、10、(14)，成等差數列；

第三部分：4、12、20、(28)，成等差數列。

故未知項為71428，正確答案為B。

解法2：將數字進行機械劃分：1|2|4、3|6|12、5|10|20、(| |)。

每個數字的三個部分構成等比數列：

1、2、4，成等比數列；

3、6、12，成等比數列；

5、10、20，成等比數列。

所以未知項的三個部分也要成等比數列，只有7、14、28符合條件，所以未知項是71428，故正確答案為B。

24. C

解析：觀察數列可知：1+1=2，1+2=3，2+3=5，3+5=8，即前兩項之和等於第三項。故空缺項為5+8=13，正確答案為C。

25. C

解析：

解析一：該數列為遞推數列。各項化可為1x2-1=1，1x2+1=3，3x2-1=5、5x2+1=11，故下一項應為11x2-1=21。故正確答案為C。

解析二：將原數列每相鄰兩個數相加，得出另外一個新數列，即2,4,8,16,？，推斷新數列的下一個數為32，反推回原數列應該是32-11=21，答案為C正確。

PART ONE
題庫練習

PART **TWO**
模擬試卷

PART THREE
考生急症室

26. B

解析：本題為遞推數列。遞推規律為：前兩項之積加上1等於下一項。具體規律為：0x1+1=1，1x1+1=2，1x2+1=3，因此原數列未知項為2x3+1=7，故正確答案為B。

27. A

解析：3x1+1=4，4x2+1=9，9x3+1=28，28x4+1=113，故未知項為113x5+1=566。故正確答案為A。

（五）Interpretation of Tables & Graphs (8 questions)

28. B

Explanation:

Average marks obtained in Physics by all the seven students

= 1/7 x [(90% of 120) + (80% of 120) + (70% of 120) + (80% of 120) + (85% of 120) + (65% of 120) + (50% of 120)]

= 1/7 x [(90 + 80 + 70 + 80 + 85 + 65 + 50)% of 120]

= 1/7 x (520% of 120)

= 624/7

= 89.14

29. B

Explanation:

From the table it is clear that Sajal and Rohit have 60% or more marks in each of the six subjects.

30. D

Explanation:

Aggregate marks obtained by Sajal = [(90% of 150) + (60% of 130) + (70% of 120)+ (70% of 100) + (90% of 60) + (70% of 40)]

= [135 + 78 + 84 + 70 + 54 + 28]

= 449

31. A

Explanation:

We shall find the overall percentage (for all the seven students) with respect to each subject. The overall percentage for any subject is equal to the average of percentages obtained by all the seven students since the maximum marks for any subject is the same for all the students. Therefore, overall percentage for:

(i) Maths: [1/7 x (90 + 100 + 90 + 80 + 80 + 70 + 65)]% = [1/7 x 575]% = 82.14%

(ii) Chemistry: [1/7 x (50 + 80 + 60 + 65 + 65 + 75 + 35)]% = [1/7 x 430]% = 61.43%

(iii) Physics: [1/7 x (90 + 80 + 70 + 80 + 85 + 65 + 50)]% = [1/7 x 520]% = 74.29%

(iv) Geography: [1/7 x (60 + 40 + 70 + 80 + 95 + 85 + 77)]% = [1/7 x 507]% = 72.43%

(v) History: [1/7 x (70 + 80 + 90 + 60 + 50 + 40 + 80)]% = [1/7 x 470]% = 67.14%

(vi) Comp. Science: [1/7 x (80 + 70 + 70 + 60 + 90 + 60 + 80)]% = [1/7 x 510]% = 72.86%

32. A

Explanation:

Percentage change (rise/fall) in the production of Company Y in comparison to the previous year, for different years are:

For 1997 = [(35-25)/25] x 100% = 40%

For 1998 = [(35-35)/25] x 100% = 0%

For 1999 = [(40-35)/35] x 100% = 14.29%

For 2000 = [(50-40)/40] x 100% = 25%

Hence, the maximum percentage rise/fall in the production of Company Y is for 1997.

PART ONE
題庫練習
PART **TWO**
模擬試卷
PART THREE
考生急症室

33. C

Explanation:

= 1/3 x (25 + 50 + 40) = 115/3 lakh tons

Average production of Company Y in the period 1998-2000

= 1/3 x (35 + 40 + 50) = 125/3 lakh tons

Therefore, required ratio = (115/3)/(125/3)

=115/125

=23/25

34. D

Explanation:

Average production (in lakh tons) in five years for the three companies are:

For Company X = 1/5 x (30 + 45 + 25 + 50 + 40) = 190/5 = 38

For Company Y = 1/5 x (25 + 35 + 35 + 40 + 50) = 185/5 = 37

For Company Z = 1/5 x (35 + 40 + 45 + 35 + 35) = 190/5 = 38

Therefore, production of five years is maximum for both the Companies X and Z.

35. A

Explanation:

The percentages of production of Company Z to the production of Company Z for various years are:

For 1996 = 35/25 x 100% = 140%

For 1997 = 40/35 x 100% = 114.29%

For 1998 = 45/35 x 100% = 128.57%

For 1999 = 35/40 x 100% = 87.5%

For 2000 = 35/50 x 100% = 70%

Clearly, this percentage is highest for 1996

能力傾向測試

模擬測驗（二）

限時四十五分鐘

PART ONE
題庫練習

PART **TWO**
模擬試卷

PART THREE
考生急症室

（一）演繹推理（8題）

請根據以下短文的內容，選出一個或一組推論。請假定短文的內容都是正確的。

1. 在一次實驗中，研究人員將大腦分為若干個區域，然後掃描並比較了每個人大腦各區域的腦灰質含量。結果顯示，智商測試中得分高的人與得分低的人相比，其大腦中有24個區域灰質含量更多，這些區域大都負責人的記憶、反應和語言等各種功能。

從這段文字中，我們可以推出：

A. 智商低的人大腦中不含灰質

B. 大腦中灰質越多的人，智商越高

C. 聰明的人在大腦24個區域中含有灰質

D. 智商高的人，記憶、反應和語言能力都強

2. 近20年來，美國女性神職人員的數量增加了兩倍多，越來越多的女性加入牧師的行列。與此同時，允許婦女擔任神職人員的宗教團體的教徒數量卻大大減少，而不允許婦女擔任神職人員的宗教團體的教徒數量則明顯增加。為了減少教徒的流失，宗教團體應當排斥女性神職人員。

如果以下陳述為真，哪一項將最有力地強化上述論證？

A. 調查顯示，77%的教徒説他們需要到教堂淨化心靈，而女性牧師在布道時卻只談社會福利問題

B. 宗教團體的教徒數量多不能説明這種宗教據有真經，所有較大的宗教在剛開始時教徒數量都很少

C.女性牧師面臨的最大壓力是神職和家庭的兼顧，有56%女性牧師說，即使有朋友幫助，也難以消除她們的憂鬱情緒

D.在允許女性擔任神職人員的宗教組織中，女性牧師很少獨立主持較大的禮拜活動

3. 意識形態：在一定的社會經濟基礎上形成的代表了某一階級或社會集團的利益的系統的思想觀念。下列屬於意識形態的是：

A. 雖然本周股市低迷，但大部分股民都相信下周股票一定會上漲

B. 某地方文學藝術界聯合會號召大家在新時期要抵制低俗文化，淨化文化環境

C. 攝影師韓先生在某知名論壇上發布了一組某藏族貧困小學生艱辛求學的照片，號召大家為其捐款，許多人紛紛回應，在網上引起了軒然大波

D. 火箭成功發射那一刻，舉國上下歡呼慶祝

4. 白領辦公室中時不時備著幾罐碳酸飲料，但有些人擔憂喝碳酸飲料會對骨骼健康造成隱患，主要原因是人體內過多的磷酸鹽會阻礙人體對鈣的吸收。

以下哪項如果為真，最不能減弱這些人的擔心？

A. 研究表明，磷酸鹽並不會對鈣的吸收產生任何負面影響

B. 研究顯示，對於大多數年齡段的人群而言，將鈣的吸收與磷酸鹽的攝取量聯繫起來是沒有必要的

C. 碳酸飲料會刺激腸胃，增加胃酸，擾亂有益菌平衡

D. 飲料當中用於增強口感的磷酸鹽只佔人體吸收磷酸鹽的極少量

5. 有經濟學者不贊成政府對低收入人群的直接救助，主張政府對大企業家和富人採取優惠的財政和稅收政策，認為大企業家和富人的投資與消費活動會促進經濟發展，增加窮人的就業機會，將財富從社會上層傳遞到社會底層。

如果以下各項為真，最能削弱上述理論的觀點是：

A. 政府對大企業家和富人採取優惠的財政和稅收政策，導致因貧富差距拉大而引發不確定的社會矛盾

B. 打破收入分配方面的平均主義，鼓勵一部分人先富起來，但具體實施這一政策存在不可預見的風險

C. 高收入者的消費需求達到飽和後，人的「自私性」將抑制投資需求，社會投資規模也不再繼續增長

D. 政府的財政和稅收政策應該優惠窮人，提高窮人購買力同樣會增加有效消費需求促使經濟持續發展

6. 時尚的一個特點是影響面廣，往往不分社會地位和社會階層，也不分男女老幼。

以下哪項如果為真，最能對上述觀點提出質疑？

A. 在歐洲君主制時代，追求奢侈是一種時尚。宮廷和貴族當然不用說了，鄉下姑娘進了城也追求奢華的生活，貴族阿 • 聖克拉拉對此感到不能容忍

B. 近來動畫片《喜羊羊和灰太狼》熱播，很多孩子都很喜歡看，有不少成年人還買了喜羊羊毛絨玩具放在沙發上或汽車裡

C. 17、18世紀的法國，男人普遍喜歡戴假髮，當然上層人和下層人的假髮的等級和價格是不一樣的

D. 有人把花上萬元去聽世界頂尖鋼琴家的演奏看作一種流行時尚，幾個退休老人說起此事都覺得不可理解

7. 為降低企業工資成本，順利度過金融危機，長江鋼廠計劃裁減一部分員工。經廠董事會討論決定，率先裁減的是那些工作效率極為低下的員工，而不是像有些企業那樣，根據工齡長短進行裁員。

長江鋼廠的這個決定最具備下列哪項前提？

A. 每個員工的工作效率是不同的

B. 工作效率低下的員工就不是好員工

C. 有能夠評判工作效率高低的可行辦法

PART ONE
題庫練習

PART TWO
模擬試卷

PART THREE
考生急症室

D. 每一個員工，尤其是那些技術工人，都會有自己的一技之長

8. 人都有趨利避害的利己主義動機，個人行為的目的總是為追求自身效用（利益）的最大化，而一件東西對某個人的效用大小取決於個人的主觀評價。在對個人效用的追求中，不僅包括對物質財富的追求，而且也包括對非物質（精神）滿足的追求。這種追求精神的滿足可以表現為扶危濟貧等利他主義行為。由此可以推出：

A. 人的利他主義行為往往出於利己主義動機

B. 人對精神滿足的追求強於對物質財富的追求

C. 扶危濟貧是人人具有的本性

D. 人人都認為物質財富越多越好

（二）Verbal Reasoning

（6 questions）

In this test, each passage is followed by three statements (the questions). You have to assume what is stated in the passage is true and decide whether the statements are either:

True (Box A): The statement is already made or implied in the passage, or follows logically from the passage.

False (Box B): The statement contradicts what is said, implied by, or follows logically from the passage.

Can't tell (Box C): There is insufficient information in the passage to establish whether the statement is true or false.

Passage 1 (Question 9 to 11)

Public Private Partnerships (PPPs) are one mode of delivery for infrastructure projects. As the name suggests, in a PPP, private contractors team up with a government department to in some way jointly deliver a piece of infrastructure. For example, the government may decide it needs a new road or hospital built. It will place it for tender. Consortiums of private companies – often consisting of a number

PART ONE
題庫練習

PART **TWO**
模擬試卷

PART THREE
考生急症室

of organisations which normally includes an investment bank, construction company, legal team, environmental consultants and equity contributors – will then bid for the project. The winning bid will then normally have the responsibility for constructing the piece of infrastructure, financing it and then operating it for a concession period during which they will receive either periodic payments from the government, or collect cash flows directly from users of the infrastructure (e.g. road tolls). PPP projects were pioneered in the United Kingdom during the late 1980s, with the then Prime Minister John Major being a key advocate. Australia was also an early adopter of the mode, with the Sydney Harbour Tunnel being the first such project to be under a PPP. The PPP method is by no means universally employed. Complete privatisation and or 100% public ownership are still far more common delivery methods.

9. The Sydney Harbour Bridge was built under a PPP.

10. The consortium of private companies usually includes a commercial bank and a construction company.

11. During the concession period, the private consortium may receive payments from the government.

Passage 2 (Question 12 to 14)

An economic bubble is a situation in which the price of an asset is greatly inflated to its intrinsic value. Famous examples of bubbles include the Dutch tulip mania of the 1600s, the Dot-com bubble of the late 1990s and early 2000s and the recent housing bubble in various countries around the world. Just like on many other economic issues, economists do not have a consensus over the causes of, effects of, or solutions to bubbles. Oft cited causes of recent bubbles include the easy/cheap availability of credit (owing in part to loose monetary policy), greater fool theory, moral hazard and herding behaviour. Most economists do agree that bubbles can lead to the misallocation of scarce resources (which is the central question of economics). Some economists believe government ought to step in to 'pop' a bubble through contractionary fiscal or monetary policy, while others believe the bubble should be allowed to run its course and deflate naturally.

12. There is consensus among economists over the solutions to bubbles.

13. Some causes of bubbles may be loose monetary policy and moral hazard.

14. The central question of economics is the allocation of scarce resources.

PART ONE
題庫練習

PART **TWO**
模擬試卷

PART THREE
考生急症室

(三) Data Sufficiency Test

(8 questions)

In this test, you are required to choose a combination of clues to solve a problem.

15. A cake recipe calls for sugar and flour in the ratio of 2 cups to 1 cup, respectively. If sugar and flour are the only ingredients in the recipe, how many cups of sugar are used when making a cake?

(1) the cake requires 33 cups of ingredients

(2) the ratio of the number of cups of flour to the total number of cups used in the recipe is 1:3

A. Statement (1) ALONE is sufficient, but statement (2) alone is not sufficient.

B. Statement (2) ALONE is sufficient, but statement (1) alone is not sufficient.

C. BOTH statements TOGETHER are sufficient, but NEITHER statement ALONE is sufficient.

D. EACH statement ALONE is sufficient.

E. Statement (1) and (2) TOGETHER are NOT sufficient to answer the question asked, and additional data are needed.

16. x is a positive integer; is x+17,283 odd?

(1) x-192,489,358,935 is odd

(2) x/4 is not an even integer

A. Statement (1) ALONE is sufficient, but statement (2) alone is not sufficient.
B. Statement (2) ALONE is sufficient, but statement (1) alone is not sufficient.
C. BOTH statements TOGETHER are sufficient, but NEITHER statement ALONE is sufficient.
D. EACH statement ALONE is sufficient.
E. Statement (1) and (2) TOGETHER are NOT sufficient to answer the question asked, and additional data are needed.

17. Chef Martha is preparing a pie for a friend's birthday. How much more of substance X does she need than substance Y?

(1) Martha needs 10 cups of substance X

(2) Martha needs the substances W, X, Y, and Z in the ratio 15:5:2:1 and she needs 4 cups of substance Y

A. Statement (1) ALONE is sufficient, but statement (2) alone is not sufficient.
B. Statement (2) ALONE is sufficient, but statement (1) alone is not sufficient.

PART ONE
題庫練習
PART TWO
模擬試卷
PART THREE
考生急症室

C. BOTH statements TOGETHER are sufficient, but NEITHER statement ALONE is sufficient.

D. EACH statement ALONE is sufficient.

E. Statement (1) and (2) TOGETHER are NOT sufficient to answer the question asked, and additional data are needed.

18. If X is a positive integer, is X divisible by 4?

(1) X has at least two 2s in its prime factorization

(2) X is divisible by 2

A. Statement (1) ALONE is sufficient, but statement (2) alone is not sufficient.

B. Statement (2) ALONE is sufficient, but statement (1) alone is not sufficient.

C. BOTH statements TOGETHER are sufficient, but NEITHER statement ALONE is sufficient.

D. EACH statement ALONE is sufficient.

E. Statement (1) and (2) TOGETHER are NOT sufficient to answer the question asked, and additional data are needed.

19. **How many prime numbers are there between the integers 7 and X, not-inclusive?**

   (1) **15<X<34**

   (2) **X is a multiple of 11 whose sum of digits is between 1 and 7**

   A. Statement (1) ALONE is sufficient, but statement (2) alone is not sufficient.
   B. Statement (2) ALONE is sufficient, but statement (1) alone is not sufficient.
   C. BOTH statements TOGETHER are sufficient, but NEITHER statement ALONE is sufficient.
   D. EACH statement ALONE is sufficient.
   E. Statement (1) and (2) TOGETHER are NOT sufficient to answer the question asked, and additional data are needed.

20. **If x is a positive integer, is x divided by 5 an odd integer?**

   (1) **x contains only odd factors**

   (2) **x is a multiple of 5**

   A. Statement (1) ALONE is sufficient, but statement (2) alone is not sufficient.
   B. Statement (2) ALONE is sufficient, but statement (1) alone is not sufficient.
   C. BOTH statements TOGETHER are sufficient, but NEI-

PART ONE
題庫練習

PART TWO
模擬試卷

PART THREE
考生急症室

THER statement ALONE is sufficient.

D. EACH statement ALONE is sufficient.

E. Statement (1) and (2) TOGETHER are NOT sufficient to answer the question asked, and additional data are needed.

21. **After a long career, John C. Walden is retiring. If there are 25 associates who contribute equally to a parting gift for John in an amount that is an integer, what is the total value of the parting gift?**

    (1) If four associates were fired for underperformance, the total value of the parting gift would have decreased by $200

    (2) The value of the parting gift is greater than $1,225 and less than $1,275

    A. Statement (1) ALONE is sufficient, but statement (2) alone is not sufficient.

    B. Statement (2) ALONE is sufficient, but statement (1) alone is not sufficient.

    C. BOTH statements TOGETHER are sufficient, but NEI-THER statement ALONE is sufficient.

    D. EACH statement ALONE is sufficient.

    E. Statement (1) and (2) TOGETHER are NOT sufficient to answer the question asked, and additional data are needed.

22. In triangle ABC, what is the measurement of angle C?

(1) The sum of the measurement of angles A and C is 120

(2) The sum of the measurement of angles A and B is 80

A. Statement (1) ALONE is sufficient, but statement (2) alone is not sufficient.

B. Statement (2) ALONE is sufficient, but statement (1) alone is not sufficient.

C. BOTH statements TOGETHER are sufficient, but NEITHER statement ALONE is sufficient.

D. EACH statement ALONE is sufficient.

E. Statement (1) and (2) TOGETHER are NOT sufficient to answer the question asked, and additional data are needed.

PART ONE
題庫練習

PART TWO
模擬試卷

PART THREE
考生急症室

（四）Numerical Reasoning

（5 questions）

Each question is a sequence of numbers with one or two numbers missing. You have to figure out the logical order of the sequence to find out the missing number(s).

(23) -1，2，1，8，19，（ ）

    A. 62

    B. 65

    C. 73

    D. 86

(24) 0，2，2，4，6，（ ）

    A. 4

    B. 6

    C. 8

    D. 10

(25) 1，2，2，5，9，（ ）

    A. 13

    B. 14

    C. 15

    D. 16

(26) 1，6，7，14，28，( )

    A. 64

    B. 56

    C. 48

    D. 36

(27) 25，36，39，46，59，( )，( )

    A. 43

    B. 58

    C. 64

    D. 73

PART ONE
題庫練習

PART TWO
模擬試卷

PART THREE
考生急症室

（五）Interpretation of Tables & Graphs (8 questions)

This is a test on reading and interpretation of data presented in tables and graphs.

Graph 1 (Question 28 to 31)

Subject	Marks out of 50				
	40 and above	30 and above	20 and above	10 and above	0 and above
Physics	9	32	80	92	100
Chemistry	4	21	66	81	100
Average (Aggregate)	7	27	73	87	100

Classification of 100 students based on the marks obtained by them in Physics and Chemistry in an examination.

28. **What is the different between the number of students passed with 30 as cut-off marks in Chemistry and those passed with 30 as cut-off marks in aggregate?**

    A. 3
    B. 4
    C. 5
    D. 6

29. If at least 60% marks in Physics are required for pursuing higher studies in Physics, how many students will be eligible to pursue higher studies in Physics?

A. 27
B. 32
C. 34
D. 41

30. The percentage of number of students getting at least 60% marks in Chemistry ove those getting at least 40% marks in aggregate, is approximately what according to the table chart?

A. 21%
B. 27%
C. 29%
D. 31%

31. The number of students scoring less than 40% marks in aggregate is?

A. 13
B. 19
C. 20
D. 27

PART ONE
題庫練習

PART **TWO**
模擬試卷

PART THREE
考生急症室

Graph 2 (Question 32 to 35)

A cosmetic company provides five different products. The sales of these five products (in lakh number of packs, lakh equals 100 000) during 1995 and 2000 are shown in the following bar graph.

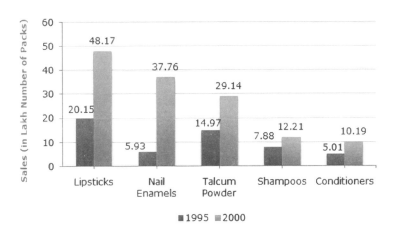

Sales (in lakh number of packs) of five different products of Cosmetic Company during 1995 and 2000

32. The sales of lipsticks in 2000 was by what percent more than the sales of nail enamels in 2000? (rounded off to nearest integer)

A. 33%

B. 31%

C. 28%

D. 22%

33. During the period 1995-2000, the minimum rate of increase in sales is in the case of?

A. Shampoos
B. Nail enamels
C. Talcum powders
D. Lipsticks

34. What is the approximate ratio of the sales of nail enamels in 2000 to the sales of Talcum powders in 1995?

A. 7:2
B. 5:2
C. 4:3
D. 2:1

35. The sales have increase by nearly 55% from 1995 to 2000 in the case of?

A. Lipsticks
B. Nail enamels
C. Talcum powders
D. Shampoos

END OF PAPER

PART ONE
題庫練習

PART TWO
模擬試卷

PART THREE
考生急症室

ANSWER - PAPER 2

（一）演繹推理（8題）

1. C

解析：智商低的人大腦中也含有灰質，只是含量少一些，所以A錯誤。大腦中與智商有關的主要是24個區域中，如果別的部分含的灰質多，與智商沒有必然聯繫，所以B不正確。智商高的人，不一定這三種能力都強，兩者之間沒有必然聯繫，所以D不正確。故答案為C。

2. A

解析：題幹觀點是「為了減少教徒的流失，宗教團體應當排斥女性神職人員」，論據是「允許婦女擔任神職人員的宗教團體的教徒數量卻大大減少，而不允許婦女擔任神職人員的宗教團體的教徒數量則明顯增加」。如果A項為真，說明女性牧師無法滿足教徒的需要，從而會導致教徒的流失，支持了題幹觀點；B、C兩項都與題幹論證無關；D項也不能加強題於論證。故答案選A。

3. B

解析：由定義可知意識形態的要點包括：在一定的社會經濟基礎上，代表某一階級或社會集團的利益，系統的思想觀念。A項、C項和D項都不是系統的思想觀念。

4. C

解析：題幹由「人體內過多的磷酸鹽會阻礙人體對鈣的吸收」得出「喝碳酸飲料會對骨骼健康造成隱患」的結論。A、B兩項直接削弱了論據，D項也能在一定程度上減弱這些人的擔心。C項說明碳酸飲料會「刺激腸胃」，不能減弱人們的擔心。故答案選C。

5. C

解析：

第一步：找出論點和論據

論點：政府不該直接救助低收入人群，而應對大企業家和富人採取優惠的財政和稅收政策。

論據：大企業家和富人的投資與消費活動會促進經濟發展，增加窮人的就業機會。

第二步：逐一判斷選項

A說的是對大企業家和富人採取優惠政策後導致的不良社會後果，跟救助低收入人群的論點無關；B說的是分配方面的平均主義，跟論點和論據無關；C證明不該給富人，D證明應該給窮人，都削弱了論點，但是D的最終結果是經濟持續發展，而不是題幹的財富傳遞，C沒有投資規模，就肯定不會有經濟增長和財富傳遞，所以C比D好。故正確答案為C。

6. D

解析：

第一步：找到論點和論據

論點：時尚的影響面廣，不分社會地位和社會階層，不分男女老幼

論據：題幹中沒有論據

第二步：判斷削弱方式

本題只有論點，沒有論據，所以需削弱論點。

第三步：逐一判斷選項

A項中，「宮廷和貴族」追求奢侈、「鄉下姑娘」也追求奢侈，說明追求奢侈這一時尚社會地位和階層，加強了論點；

B項中說明喜羊羊這一動畫形像受到人們的喜愛，且不分老幼，加強了論點；

C項說明17、18世紀的法國，男人喜歡戴假髮，這一潮流是不分階層的，加強了論點；

D項「幾個退休老人」不能理解有些人花高價去聽演奏，說明「聽世界頂尖鋼琴家的演奏」這種時尚在不同的年齡層中存在分歧，通過舉反例削弱論點。故正確答案選D。

PART ONE
題庫練習

PART TWO
模擬試卷

PART THREE
考生急症室

7. C

解析：

第一步：找出論點和論據

本題論點為董事會討論決定率先裁減工作效率極為低下的員工。沒有論據。

第二步：判斷加強方式

論據有利於說明論點，本題只有論點沒有論據，所以需要增加論據加強論點。

第三步：逐一判斷選項

C項能夠加強論點，如果董事會沒有能夠評判工作效率高低的可行辦法，就無法判斷哪些員工工作效率極為低下，也就無法率先裁剪工作效率極為低下的員工，故C項是董事會作出該決定的前提。A項員工的工作效率是否相同、B項工作效率低下的員工是不是好員工、D項員工是否有一技之長更董事會作出的裁員決定都沒有直接關係，屬於無關選項。故正確答案為C。

8. A

解析：本題考查的是結論類中的主題一致原則。由題幹的第二句話可知人的利他主義源於對自身效用的追求。因此，本題答案為A選項。

（二）Verbal Reasoning (6 questions)

9. C (Can't tell)

Explanation: Read the question/passage very carefully. The passage says that The Sydney Harbour Tunnel was built under a PPP, but there is no mention of The Sydney Harbour Bridge being built under a PPP. It's not explicitly ruled out either though, and so the answer must be Can't Tell.

10. C (Can't tell)

Explanation: There is mention of a construction company being in the consortium, but no mention of a commercial bank. A commercial bank's presence wasn't ruled out, though, because of the phrase …'normally includes…'. Again this is a question to read very carefully, and it is set up to trick you, because there is mention of an investment bank being part of the consortium normally.

11. A (True)

Explanation: The 4th last sentence says this.

12. B (False)

Explanation: Sentence 3 contradicts this directly– there is no consensus.

13. A (True)

Explanation: Sentence 4 includes both these in possible causes.

14. A (True)

Explanation: Penultimate sentence says this.

(三) Data Sufficiency Test (8 questions)

15. A

Explanation: 1. Based upon the question, we can set up a few equations:

Equation (1):Sugar/Flour=2/1

Since one cake could be made from 2 cups of sugar and 1 cup of flour (or different number of cups in the same ratio):

Equation (2): Sugar/(Total Ingredients)=2/(2+1)=2/3

Equation (3): Flour/(Total Ingredients)=1/(2+1)=1/3

2. Evaluate Statement (1) alone:

(a) Since the cake requires 33 cups of ingredients, using Equation (2), we know that Total Ingredients=33:

Sugar/(Total Ingredients)=2/3

Sugar/33=2/3

Therefore: Sugar=22 cups

(b) Statement (1) is SUFFICIENT.

PART ONE
題庫練習

PART TWO
模擬試卷

PART THREE
考生急症室

3. Evaluate Statement (2) alone:

(a) Statement (2) does not provide any new information. Based upon the original question, we derived Equation (3). Statement (2) is merely a restatement of Equation (3).

(b) Consider two examples:

If there were 10 cups of flour, the total amount of ingredients would be 30 cups and there would be 20 cups of sugar.

But, if there were 5 cups of flour, the total amount of ingredients would be 15 cups and there would be 10 cups of sugar.

(c) Statement (2) is NOT SUFFICIENT since we cannot determine how many cups of sugar were used in the cake.

4. Since Statement (1) alone is SUFFICIENT but Statement (2) alone is NOT SUFFICIENT, answer A is correct.

(16) A

Explanation: 1. Before evaluating Statements (1) and (2), it is extremely helpful to keep in mind that an odd number is the result of a sum of numbers with unlike parity. In other words: even+odd=odd. Since 17,283 is odd, the only way x+17,283 will be odd is if x is even. Consequently, the simplified version of the question is: is x even?

2. Evaluate Statement (1) alone.

(a) Statement (1) says that x-192,489,358,935 is odd. Since there is only one way for a difference to be odd (i.e., if the parity of the numbers is different), Statement (1) implies that x is even (otherwise, if x were odd, x-192,489,358,935 would be even). Since Statement (1) gives the parity of x, it is SUFFICIENT.

(b) Statement (1) alone is SUFFICIENT.

3. Evaluate Statement (2) alone:

(a) Statement (2) says that x/4 is not an even integer. It is important to note that this does not mean that x cannot be even (e.g., 6 is even yet 6/4 is not an even integer). Possible values of x include 2, 3, 6, 10, 11.

x	Parity of x	x/4
2	Even	Not an even integer

3	Odd	Not an even integer
6	Even	Not an even integer
10	Even	Not an even integer
11	Odd	Not an even integer

(b) As this list indicates, there is no definitive information about the parity of x (e.g., 11 is odd and 10 is even). Consequently, Statement (2) is NOT SUFFICIENT.

(c) Statement (2) alone is NOT SUFFICIENT.

4. Since Statement (1) alone is SUFFICIENT and Statement (2) alone is NOT SUFFICIENT, answer A is correct.

17. B

Explanation: 1. The phrase "how much more of substance X does she need than substance Y" can be translated into algebra as: X-Y

2. Evaluate Statement (1) alone:

(a) Although X=10, there is no information about Y. Consequently, we cannot determine the value of X-Y

(b) Statement (1) is NOT SUFFICIENT.

3. Evaluate Statement (2) alone:

(a) Y=4 and the ratio of X:Y=5:2. Consequently, X=10 and X-Y=10-4=6

(b) Statement (2) is SUFFICIENT.

4. Since Statement (1) alone is NOT SUFFICIENT but Statement (2) alone is SUFFICIENT, answer B is correct.

18. A

Explanation: 1. A number is divisible by any of its prime factors or any combination of its prime factors. For a number to be divisible by four, it must have two 2s in its prime factorization since 2x2=4 and, if 4 is a factor of X, X will be divisible by 4.

2. Evaluate Statement (1) alone:

(a) Since X has two 2s in its prime factorization, 4 must be a factor of X and, consequently, X must be divisible by 4. Statement (1) is SUFFICIENT.

(b) If this seems too abstract, consider the following examples which show

PART ONE
題庫練習

PART **TWO**
模擬試卷

PART THREE
考生急症室

that whenever X has at least two 2s in its prime factorization (which it must as per Statement (1)), X is divisible by 4:

X=4: has two 2s in its prime factorization and, as a result, is divisible by 4

X=6: has only one 2 in its prime factorization and, as a result, is not divisible by 4

X=8: has at least two 2s in its prime factorization and, as a result, is divisible by 4

X=10: has only one 2 in its prime factorization and, as a result, is not divisible by 4

(c) Since X cannot be 6, 10, et   C. as these values do not have at least two 2s as prime factors (as is required by Statement (1)), X can only be 4, 8, et   C. and will always be divisible by 4.

(d) Statement (1) alone is SUFFICIENT.

3. Evaluate Statement (2) alone:

(a) If X is divisible by 2, you know that X must have at least one 2 in its prime factorization. However, you do not know that X has two 2s in its prime factorization and, as a result, you cannot be sure that X is divisible by 4.

(b) If this seems too abstract, consider the following examples, all of which are divisible by 2 in keeping with the requirements of Statement (2):

X=4: has two 2s in its prime factorization and, as a result, is divisible by 4

X=6: has only one 2 in its prime factorization and, as a result, is not divisible by 4

X=8: has at least two 2s in its prime factorization and, as a result, is divisible by 4

X=10: has only one 2 in its prime factorization and, as a result, is not divisible by 4

(c) Statement (2) alone is NOT SUFFICIENT.

4. Since Statement (1) alone is SUFFICIENT and Statement (2) alone is NOT SUFFICIENT, answer A is correct.

19. E

Explanation: 1. In evaluating this problem, it is important to keep in mind the list of possible prime numbers: 7, 11, 13, 17, 19, 23, 29, 31, 37, 41, 43,

47, 53

2. Evaluate Statement (1) alone:

(a) The prime numbers between 15 and 34, not-inclusive, include: 17, 19, 23, 29, 31

(b) Since there is no definitive information about the value of X, we do not know how many prime numbers exist between 7 and X.

If X=17, there would be 2 prime numbers between 7 and X (i.e., 11 and 13).

If X=18, there would be 3 prime numbers between 7 and X (i.e., 11, 13, and 17).

If X=21, there would be 4 prime numbers between 7 and X (i.e., 11, 13, 17, and 19).

There is not enough information to definitively answer the question.

(c) Statement (1) alone is NOT SUFFICIENT.

3. Evaluate Statement (2) alone:

(a) List the multiples of 11 and their sums (stopping when the sum is no longer less than 7).

x=11; sum of digits is 1+1=2

x=22; sum of digits is 2+2=4

x=33; sum of digits is 3+3=6

x=44; sum of digits is 4+4=8, which is too high so x cannot be greater than 33.

(b) Since X can be 11, 22, or 33, there are different possible answers to the question of how many prime numbers are there between the integers 7 and X:

If X=11, there would be 0 prime numbers between7 and X.

If X=22, there would be 4 prime numbers between7 and X (i.e., 11, 13, 17, and 19).

There is not enough information to definitively answer the question.

(c) Statement (2) alone is NOT SUFFICIENT.

4. Evaluate Statements (1) and (2) together:

(a) Putting Statements (1) and (2) together, X must meet the following conditions:

PART ONE
題庫練習
PART **TWO**
模擬試卷
PART THREE
考生急症室

(1) 15<X<34

(2) X=11, 22, 33

This means that possible values for X include: X=22 or 33

(b) The two possible values for X give different answers to the original question:

If X=22, there would be 4 prime numbers between 7 and X (i.e., 11, 13, 17, and 19).

If X=33, there would be 7 prime numbers between 7 and X (i.e., 11, 13, 17, 19, 23, 29, and 31).

(c) Statements (1) and (2), even when taken together, are NOT SUFFICIENT.

5. Since Statement (1) alone is NOT SUFFICIENT, Statement (2) alone is NOT SUFFICIENT, and Statements (1) and (2), even when taken together, are NOT SUFFICIENT, answer E is correct.

20. C

1. A number divided by 5 will be an odd integer if and only if that number contains only odd factors, one of which is 5. In other words, there are two conditions under which x divided by 5 will be an odd integer:

(1) x is a multiple of 5

(2) x contains only odd factors

2. Evaluate Statement (1) alone:

(a) If x contains only odd factors, there is no guarantee that one of those factors is 5. Consequently, there is no guarantee that x will be divisible by 5.

(b) For example:

9=3x3→not an odd integer when divided by 5 since 5 is not a factor

21=3x7→not an odd integer when divided by 5 since 5 is not a factor

15=3x5→an odd integer when divided by 5 since 5 is a factor

105=3x7x5→an odd integer when divided by 5 since 5 is a factor

(c) Statement (1) alone is NOT SUFFICIENT.

3. Evaluate Statement (2) alone:

(1) Simply because x is a multiple of 5 does not guarantee that x only con-

tains odd factors. Consequently, there is no guarantee that x is divisible by 5.

5=5x1→an odd integer when divided by 5 because 5 is a factor and there are only odd factors

15=5x3→an odd integer when divided by 5 because 5 is a factor and there are only odd factors

25=5x5→an odd integer when divided by 5 because 5 is a factor and there are only odd factors

20=5x4→not an odd integer when divided by 5 because there is at least one even factor

30=6x5→not an odd integer when divided by 5 because there is at least one even factor

(2) Statement (2) alone is NOT SUFFICIENT.

4. Evaluate Statements (1) and (2) together:

(a) With only odd factors and x as a multiple of 5 (i.e., with 5 as a factor), you know that x divided by 5 must be an odd number since the two conditions laid out earlier are fulfilled.

(b) Consider the following examples:

5=5x1→an odd integer when divided by 5

15=5x3→an odd integer when divided by 5

25=5x5→an odd integer when divided by 5

35=5x7→an odd integer when divided by 5

(c) Statements (1) and (2), when taken together, are SUFFICIENT.

5. Since Statement (1) alone is NOT SUFFICIENT and Statement (2) alone is NOT SUFFICIENT, but Statements (1) and (2), when taken together, are SUFFICIENT, answer C is correct.

21. D

1. Simplify the question by translating it into algebra.

Let P=the total value of John's parting gift

Let E=the amount each associate contributed

Let N=the number of associates

P=NE=25E

PART ONE
題庫練習
PART **TWO**
模擬試卷
PART THREE
考生急症室

2. With this algebraic equation, if you find the value of either P or E, you will know the total value of the parting gift.

3. Evaluate Statement (1) alone:

(a) Two common ways to evaluate Statement (1) alone:

4. Statement 1: Method 1:

(a) Since the question stated that each person contributed equally, if losing four associates decreased the total value of the parting gift by $200, then the value of each associate's contribution was $50(=$200/4).

(b) Consequently, P=25E=25(50)=$1,250

5. Statement 1: Method 2

(a) If four associates leave, there are N-4=25-4=21 associates.

(b) If the value of the parting gift decreases by $200, its new value will be P-200.

(c) Taken together, Statement (1) can be translated:

P-200=21E

P=21E+200

(d) You now have two unique equations and two variables, which means that Statement (1) is SUFFICIENT.

(e) Although you should not spend time finding the solution on the test, here is the solution.

Equation 1: P=21E+200

Equation 2: P=25E

P=P

25E=21E+200

4E=200

E=$50

(f) P=NE=25E=25($50)=$1250

6. Evaluate Statement (2) alone:

(a) Statement (2) says that $1,225<P<$1,275. It is crucial to remember that the question stated that "25 associates contribute equally to a parting gift for John in an amount that is an integer." In other words P/25 must be an integer. Stated differently, P must be a multiple of 25.

(b) There is only one multiple of 25 between 1,225 and 1,275. That number is $1,250. Since there is only one possible value for P, Statement (2) is SUFFICIENT.

7. Since Statement (1) alone is SUFFICIENT and Statement (2) alone is SUFFICIENT, answer D is correct.

22. B

Explanation: 1. Since the sum of the measure of the interior angles of a triangle equals 180 degrees, you can write the following equation: The measure of angles A+B+C=180

2. Evaluate Statement (1) alone:

(a) Translate Statement (1) into algebra: A+C=120

(b) Use the foundational triangle equation (i.e., all angles add up to 180):

A+B+C=180

(A+C)+B=180

Substitute A+C=120 into the equation. 120+B=180, B=60

(c) It is impossible to determine the value of angle C. Angle A could be 60 degrees and angle C could be 60 degrees. However, angle A could be 20 degrees and angle C could be 100 degrees.

(d) Statement (1) is NOT SUFFICIENT.

3. Evaluate Statement (2) alone:

(1) Translate Statement (2) into algebra:

A+B=80

(2) Use the foundational triangle equation (i.e., all angles add up to 180):

A+B+C=180

(A+B)+C=180

Substitute A+B=80 into the equation.

80+C=180

C=100

(3) Statement (2) is SUFFICIENT.

4. Since Statement (1) alone is NOT SUFFICIENT and Statement (2) alone is SUFFICIENT, answer B is correct.

PART ONE
題庫練習

PART TWO
模擬試卷

PART THREE
考生急症室

（四）Numerical Reasoning (5 questions)

23. A

解析：本題為遞推數列。遞推規律為：第一項的3倍加上第二項的2倍得到第三項。具體規律為：(-1)x3+2x2=1，2x3+1x2=8，1x3+8x2=19，因此原數列的下一項為：8x3+19x2=62，故正確答案為A。

24. D

解析：前兩項之和等於後一項。故選D。

25. D

解析：前三項之和等於下一項。故選D。

26. B

解析：所有前項的和等於下一項。故( )=1+6+7+14+28=56。故選B。

27. C、D

解析：相鄰三項和構成平方數列，即：25+36+39=100、36+39+46=121、39+46+59=144、46+59+(64)=169、59+(64)+(73)=196。故選C、D。

（五）Interpretation of Tables & Graphs (8 questions)

28. D

Explanation:

Required difference from the table chart

= (No. of students scoring 30 and above marks in Chemistry) - (Number of students scoring 30 and above marks in aggregate)

= 27 - 21

= 6

29. B

Explanation:

We have 60% of 50 = (60/100) x 50 = 30

Therefore Required number

= No. of students scoring 30 and above marks in Physics

= 32

30. C

Explanation:

Number of students getting at least 60% marks in Chemistry

= Number of students getting 30 and above marks in Chemistry

= 21

Number of students getting at least 40% marks in aggregate

= Number of students getting 20 and above marks in aggregate

= 73

Required percentage = (21/73)x100%

= 28.77%

= 29%

31. D

Explanation:

We have 40% of 50 = (40/100) x 50 = 20

Therefore, required number

= Number of students scoring less than 20 marks in aggreagate

= 100 - Number of students scoring 20 and above marks in aggregate

= 100 - 73

= 27

32. C

Explanation:

required percentage = [(48.14 - 37.76)/ 37.76] x 100%

PART ONE
題庫練習
PART TWO
模擬試卷
PART THREE
考生急症室

= 27.57%

= approximately 28%

33. A

Explanation:

The percentage increase from 1995 to 2000 for various products are:

Lipsticks = [(48.17 - 20.15) / 20.15] x 100% = 139.06%

Nail enamels = [(37.76 - 5.93) / 5.93] x 100% = 536.76%

Talcum powders = [(29.14 - 14.97) / 14.97] x 100% = 94.66%

Shampoos = [(12.21 - 7.88) / 7.88] x 100% = 54.95%

Conditioners = [(10.19 - 5.01) / 5.01] x 100% = 103.39%

Therefore, the minimum rate of increase in sales from 1995 to 2000 is in the case of Shampoos.

34. B

Explanation:

Required ratio:

= 37.76/14.97

~ = 2.5

= 5/2

35. D

Explanation:

The percentage increase from 1995 to 2000 for various products are:

Lipsticks = [(48.17 - 20.15) / 20.15] x 100% = 139.06%

Nail enamels = [(37.76 - 5.93) / 5.93] x 100% = 536.76%

Talcum powders = [(29.14 - 14.97) / 14.97] x 100% = 94.66%

Shampoos = [(12.21 - 7.88) / 7.88] x 100% = 54.95% (approximately 55%)

Conditioners = [(10.19 - 5.01) / 5.01] x 100% = 103.39%

考生急症室一

1） 每隔多久考CRE一次？

CRE一年考兩次，分別在6月和10月考試。

2） 什麼人符合申請資格？

- 持有大學學位（不包括副學士學位）；或
- 現正就讀學士學位課程最後一年；或
- 持有符合申請學位或專業程度公務員職位所需的專業資格。

3） 「綜合招聘考試」(CRE)跟「聯合招聘考試」(JRE)有何分別？

在CRE中英文運用考試中取得「二級」成績後，可投考JRE，考試為AO、EO及勞工事務主任、貿易主任四職系的招聘而設。

4） 可否使用CRE的成績來申請政府以外的工作？

CRE招聘考試是為招聘學位或專業程度公務員職位而設的基本測試，而非一項學歷資格。

PART ONE
題庫練習

PART TWO
模擬試卷

PART THREE
考生急症室

5) 如遺失了CRE考試／基本法測試的成績通知書，可否申請補領？

可以書面（地址：香港添馬添美道2號政府總部西翼7樓718室）或電郵形式（電郵地址：csbcseu@csb.gov.hk）向公務員考試組提出申請。

看得喜 放不低

創出喜閱新思維

書名	公務員招聘 能力傾向測試 精讀王NOTE
ISBN	978-988-74806-4-8
定價	HK$128
出版日期	2020年11月
作者	Fong Sir
責任編輯	鄭浩文
編審	香港通識教育資源及創新協會
版面設計	梁文俊
出版	文化會社有限公司
電郵	editor@culturecross.com
網址	www.culturecross.com
發行	香港聯合書刊物流有限公司
	地址：香港新界大埔汀麗路36號中華商務印刷大廈3樓
	電話：（852）2150 2100
	傳真：（852）2407 3062

網上購買 請登入以下網址：

一本 My Book One　　　超閱網 Superbookcity　　　香港書城 Hong Kong Book City

www.mybookone.com.hk　　www.mybookone.com.hk　　www.hkbookcity.com